Electrostatic Discharge - From Electrical breakdown in Micro-gaps to Nano-generators

Edited by Steven H. Voldman

Published in London, United Kingdom

IntechOpen

Supporting open minds since 2005

Electrostatic Discharge - From Electrical breakdown in Micro-gaps to Nano-generators
http://dx.doi.org/10.5772/intechopen.81456
Edited by Steven H. Voldman

Contributors
Chun-Yu Lin, He Zhang, Liwei Quan, Jie Wang, Di Liu, Linglin Zhou, Zhong Lin Wang, Guodong Meng, Yonghong Cheng, Steven Voldman

First published in London, United Kingdom, 2019 by IntechOpen
IntechOpen is the global imprint of INTECHOPEN LIMITED, registered in England and Wales, registration number: 11086078, The Shard, 25th floor, 32 London Bridge Street
London, SE19SG - United Kingdom
Printed in Croatia

British Library Cataloguing-in-Publication Data
A catalogue record for this book is available from the British Library

Additional hard and PDF copies can be obtained from orders@intechopen.com

Electrostatic Discharge - From Electrical breakdown in Micro-gaps to Nano-generators
Edited by Steven H. Voldman
p. cm.
Print ISBN 978-1-78984-896-0
Online ISBN 978-1-78984-897-7
eBook (PDF) ISBN 978-1-83962-265-6

For EU product safety concerns:
IN TECH d.o.o., Prolaz Marije Krucifikse Kozulić 3, 51000 Rijeka, Croatia, info@intechopen.com or visit our website at intechopen.com.

Meet the editor

Dr. Steven H. Voldman is the first IEEE Fellow in the field of electrostatic discharge (ESD) for "Contributions in ESD protection in CMOS, SOI, and SiGe Technology." He received his degrees from the University of Buffalo, Massachusetts Institute of Technology (MIT), and the University of Vermont.

Dr. Voldman was a member of development teams at IBM, the Taiwan Semiconductor Manufacturing Corporation (TSMC), and Samsung. In 2018 he served as an International Visiting Scholar at Khon Kaen University in Thailand. Dr. Voldman is the author of a ten-book series and 150 technical papers, and recipient of 262 issued US patents. He initiated a university lecture program to Thailand, Malaysia, Singapore, United States, Korea, Taiwan, Philippines, Sri Lanka, Senegal, Swaziland, Zimbabwe, Nepal, and China.

Contents

Preface **XI**

Section 1
Electrostatic Design in CMOS **1**

Chapter 1 **3**
Introductory Chapter: Electrostatic Discharge Breakdown in Micro-gaps and Nanogenerators
by Steven Voldman

Chapter 2 **9**
Low-C ESD Protection Design in CMOS Technology
by Chun-Yu Lin

Section 2
Breakdown in Micro-gaps **27**

Chapter 3 **29**
Electrical Breakdown Behaviors in Microgaps
by Guodong Meng and Yonghong Cheng

Section 3
Nanogenerators **49**

Chapter 4 **51**
Nanogenerators from Electrical Discharge
by Jie Wang, Di Liu, Linglin Zhou and Zhong Lin Wang

Chapter 5 **69**
Theoretical Prediction and Optimization Approach to Triboelectric Nanogenerator
by He Zhang and Liwei Quan

Preface

As we enter the nanoelectronics era, electrostatic discharge (ESD) phenomena is still an important issue for everything from micro-electronics to nanostructures. One of the driving forces for this interest is the concern of the ability to manufacture nanostructures without destruction associated with static charge and ESD events.

The book *Electrostatic Discharge: From Electrical Breakdown in Micro-gaps to Nanogenerators* opens with an introductory chapter. The first section of this book discusses the issue of ESD in advanced technologies. The second section follows with a chapter on electrical breakdown in micro-gaps. The third section contains a chapter on nanogenerators from ESD and a chapter on theoretical prediction and optimization of triboelectric nanogenerators.

This book provides insight into the operation and design of micro-gaps and nanogenerators. It will be of use for engineers and scientists that have an interest in ESD physics and design.

Dr. Steven H Voldman
Steven H Voldman LLC,
Lake Placid,
New York, USA

Section 1

Electrostatic Design in CMOS

Chapter 1

Introductory Chapter: Electrostatic Discharge Breakdown in Micro-gaps and Nanogenerators

Steven Voldman

1. Introduction

As we enter the nanoelectronic era, electrostatic discharge (ESD) phenomena is still an important issue for micro-electronics and nanostructures [1–10]. This interest is the concern of the ability to manufacture nanostructures without destruction associated with static charge and ESD events. ESD issues are a concern with almost all structures with a small gap in the device. The introduction will provide examples of structures that have ESD concerns which will establish a base understanding for the future chapters in this book.

2. Photomasks

Photomasks, or "masks," are used for exposing the photosensitive materials for designing components. These photomasks must be "defect free." Shapes are formed on a glass insulating surface to form the masks, where the shapes are metallic

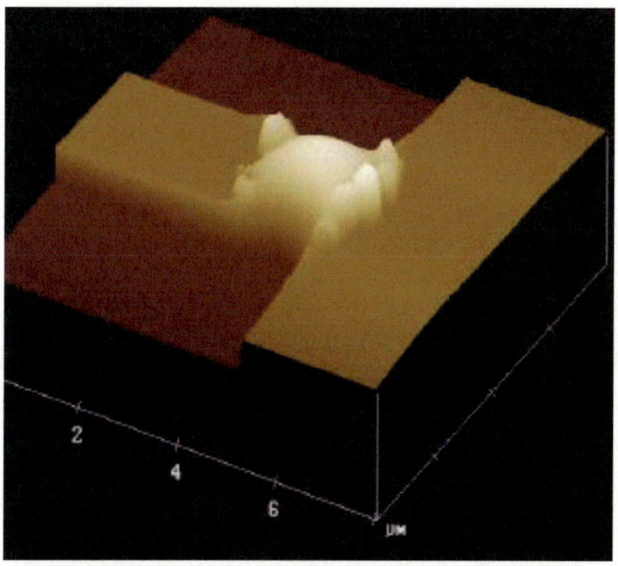

Figure 1.
Atomic force microscopy image of photomask ESD damage.

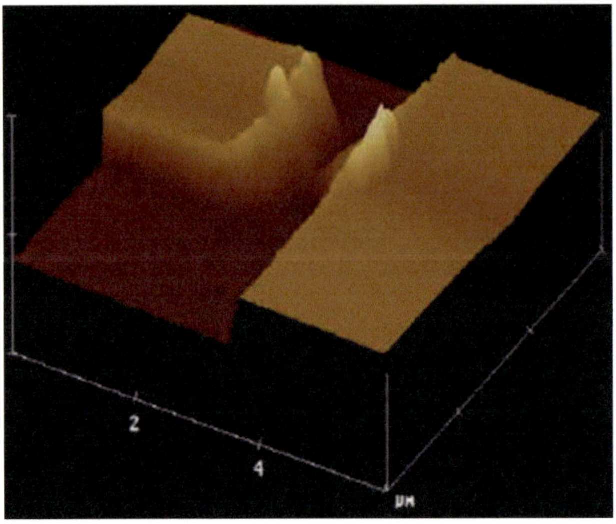

Figure 2.
Atomic force microscopy image of latent photomask ESD damage.

chrome. A key problem is the buildup of charge on the mask shapes [5]. Between each chrome shape on the mask, a potential "micro-gap" exists which can lead to electrical discharge when the electric potential exceeds the air breakdown. The law that governs the breakdown is known as Paschen's law. For gases, Paschen's law states for electrical breakdown of gases, the breakdown is a function of the product of gas pressure and gap width.

With the dimensional scaling, smaller line width, and the spacing between lines also are reduced, leading to electrostatic micro-discharges occurring between the mask shapes. **Figure 1** shows an example of an ESD discharge as a function of the spacing between two shapes on a photomask.

The mask shape damage can introduce defects in the product chip. With dimensional scaling, the spacing decreases.

Figure 2 shows a second image of a mask damage. These "nano-defects" can also lead to defects in the product chip.

3. Magnetic recording

In the magnetic recording industry, a small thin film magneto-resistor (MR) is used [6]. To sense the magnetic field, a magneto-resistor is mounted on a "magnetic head." To continue to scale down the size of disk, the magnetic recording industry continues to evolve to new devices. The devices are MR heads, to the giant magneto-resistors (GMR) and tunneling magneto-resistor (TMR).

Figure 3 shows damage in a MR structure. The damaged device leads to dimensional changes in the MR stripe, causing a change in series resistance of the MR stripe. Micro-breakdown can also occur between the magneto-resistor, the adjacent shields, and substrates. Along the surface, breakdown can occur leading to damage of the MR stripe and the physical surface.

With technology scaling, the size and film thickness are reduced to sense smaller signals; as a result, the human body model (HBM) ESD "robustness" is decreased from 150 and 35 to less than 10 V.

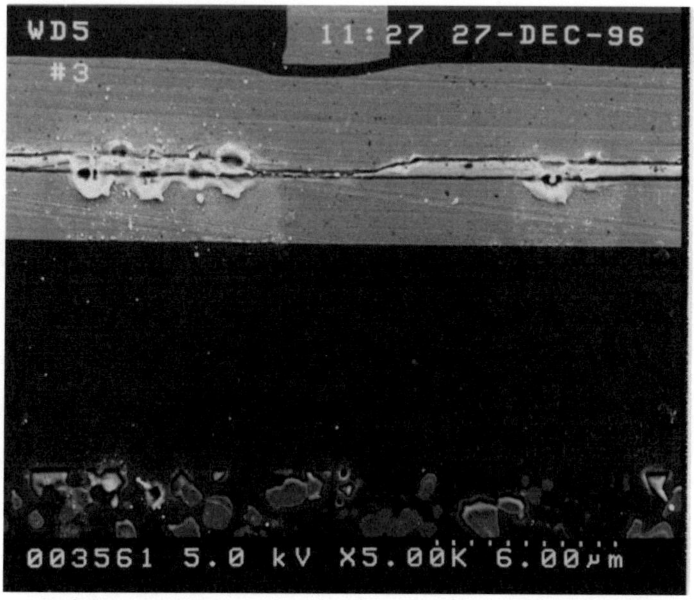

Figure 3.
Magnetic recording head after ESD damage.

4. FinFET transistors

In semiconductors, to pack more transistors into a smaller space and to provide higher performance, the transistor is leaving the two-dimensional wafer and becoming more three-dimensional. Today, in advanced technologies, these new devices are known as "FinFETs." The FinFET is a multi-finger structure. The MOSFET gate covers the "fin" over the parallel fin structures. The FinFET consists of nano-channels to conduct the MOSFET current [7].

5. MEMS

Microelectromechanical systems (MEMS) are being designed for motors, generator, micro-mirrors, switches, and passive circuit elements [9, 10]. MEMs pose a new challenge due to many of these nano-elements are electrostatically actuated elements and contain micro-gaps.

In MEM structures, segments of the elements are closely spaced and inherently have micro-gaps. Electrical spark can occur in the gap leading to melting of the component and "stiction."

In micro-motors, this issue also occurs [8]. It is noted, in micro-motors the existence of damage to the gear rotation, and "nano-welding" was observed from the current of the ESD event.

For RF applications, RF MEM switches have advantages compared to conventional switches. But, broken physical elements can lead to residual materials within the air gap, influencing functional operation or electrical shorting.

ESD damage occurs in the RF switch between the input and output [9], as well as between the actuator and the switch input and output. **Figure 4** shows an example of the damage observed in the RF switch.

Micro-mirrors have present day and future applications in the system. ESD events can lead to damage between the mirror and actuator; this can lead to

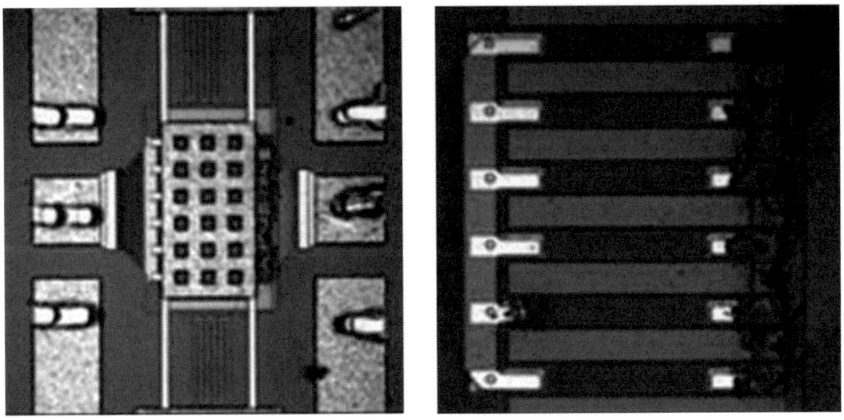

Figure 4.
RF switch damage after ESD testing (with and without membrane removed).

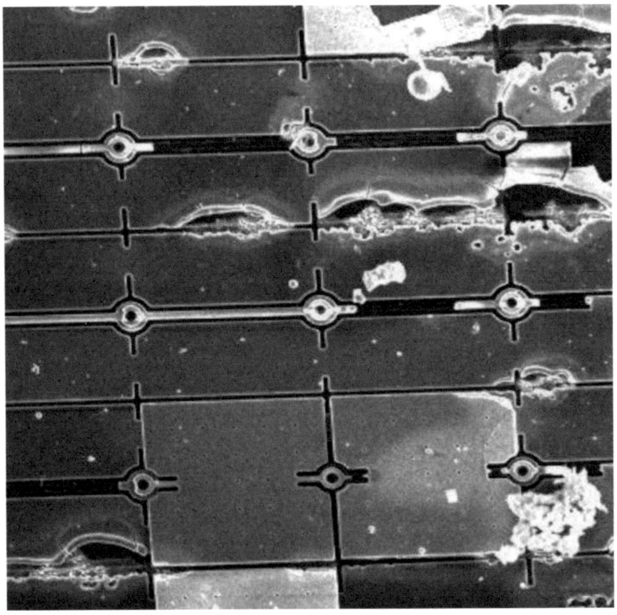

Figure 5.
Micro-mirror array damage after ESD testing.

operational issues affecting the tilt angle and rotation of the micro-mirror structures. As we move to nano-gaps and spaces, these concerns will continue to exist. **Figure 5** shows damage to the micro-mirrors after ESD testing.

6. Closing comments and summary

As nanostructures become smaller, the electrostatic sensitivity increases leading to new failure mechanisms. Electrostatically actuated devices, magnetic recording devices, photomasks, RF switches, and micromachines have air gaps which can lead to breakdown across surfaces and in the air gap.

This book will provide insight into operation and design of micro-gaps and nano-generators. It will be enlightening for engineers and scientists that have an interest in electrostatic discharge physics and design.

Author details

Steven Voldman
LLC, New York, USA

*Address all correspondence to: voldman@ieee.org

References

[1] Voldman S. Lightning rods for nanoelectronics. Scientific American. 2002;**287**(4):90-97

[2] Voldman S. Electrostatic discharge protection in the nano-technology—Will we be able to provide ESD protection in the future? Invited Talk. In: Proceedings of the International Conference on Semiconductors and Integrated Circuit Technology (ICSICT); Shanghai, China; 2006

[3] Voldman S. Electrostatic discharge in nano-technology. In: Keynote Talk, Application Specific Circuits and Networks (ASICON) 2007; Guilin, China; October 12-15, 2007

[4] Voldman S. ESD: Failure Mechanisms and Models. Chichester, England: John Wiley and Sons, Ltd; 2009

[5] Montoya J, Levit L, Englisch A. A study of the mechanisms for ESD damage in reticles. In: Proceedings of the Electrical Overstress/Electrostatic Discharge (EOS/ESD) Symposium; 2000. pp. 394-405

[6] Chen TW, Wallash AJ, Dutton R. Ultra-fast transmission line pulse testing of tunneling and giant magnetoresistor heads. In: Proceedings of the Electrical Overstress/Electrostatic Discharge (EOS/ESD) Symposium; 2008. pp. 258-261

[7] Russ C, Gossner H, Schulz T, Chaudhary N, Xiong W, Marshall A, et al. ESD evaluation of emerging MUGFET technology. In: Proceedings of the Electrical Overstress/Electrostatic Discharge (EOS/ESD) Symposium; 2005. pp. 280-289

[8] Walraven JA, Soden JM, Cole EI, Tanner DM, Anderson RE. Human body model, machine model, and charged device model ESD testing of surface micromachined microelectromechanical systems (MEMS). In: Proceedings of the Electrical Overstress/Electrostatic Discharge (EOS/ESD) Symposium; 2001. pp. 238-247

[9] Tazzoli A, Peretti V, Zanoni E, Meneghesso G. Transmission line pulse (TLP) testing of radio frequency (RF) micro-machined micro-electromechanical systems (MEMS). In: Proceedings of the Electrical Overstress/Electrostatic Discharge (EOS/ESD) Symposium; 2006. pp. 295-303

[10] Sangameswaran S, De Coster J, Linten D, Scholz M, Thijs S, Haspeslagh L, et al. ESD reliability issues in microelectromechanical systems (MEMS): A case study of micromirrors. In: Proceedings of the Electrical Overstress/Electrostatic Discharge (EOS/ESD) Symposium; 2008. pp. 249-257

Chapter 2

Low-C ESD Protection Design in CMOS Technology

Chun-Yu Lin

Abstract

Electrostatic discharge (ESD) protection design is needed for integrated circuits in CMOS technology. The choice for ESD protection devices in the CMOS technology includes diode, MOSFET, and silicon controlled rectifier (SCR). These ESD protection devices cause signal losses at high-frequency input/output (I/O) pads due to the parasitic capacitance. To minimize the impacts from ESD protection circuit on high-frequency performances, ESD protection circuit at I/O pads must be carefully designed. A review on ESD protection designs with low parasitic capacitance for high-frequency applications in CMOS technology is presented in this chapter. With the reduced parasitic capacitance, ESD protection circuit can be easily combined or co-designed with high-frequency circuits. As the operating frequencies of high-frequency circuits increase, on-chip ESD protection designs for high-frequency applications will continuously be an important design task.

Keywords: CMOS, ESD protection, high frequency, high speed, low capacitance

1. Introduction

The integrated circuits (ICs) operated at higher frequency are needed. For example, the transceivers operated in gigahertz (GHz) bands are the good candidate for the demand of faster data transmission [1]. CMOS technology is a promising way to implement the GHz integrated circuits with the advantages of high integration capability and low cost for mass production [2, 3]. However, the transistors in CMOS and even FinFET technologies are inherently susceptible to the electrostatic discharge (ESD) events [4, 5]. Once any transistor is damaged by ESD, it cannot be recovered, and the IC functionality will be lost. Therefore, the ESD protection design must be equipped on the chip. Nevertheless, the ESD protection devices cause the IC performance degradation. The ICs operated in GHz frequencies are very sensitive to the parasitic capacitance [6, 7]. To mitigate the performance degradation caused by ESD protection device, the low-capacitance (low-C) ESD protection designs are needed [8, 9].

2. ESD protection requirement

To adequately protect the ICs, the ESD protection circuit must shunt ESD current with limited voltage drop [10–12]. **Figure 1** shows the ESD design window of an IC, which is defined by the power-supply voltage (V_{DD} and V_{SS}) and the

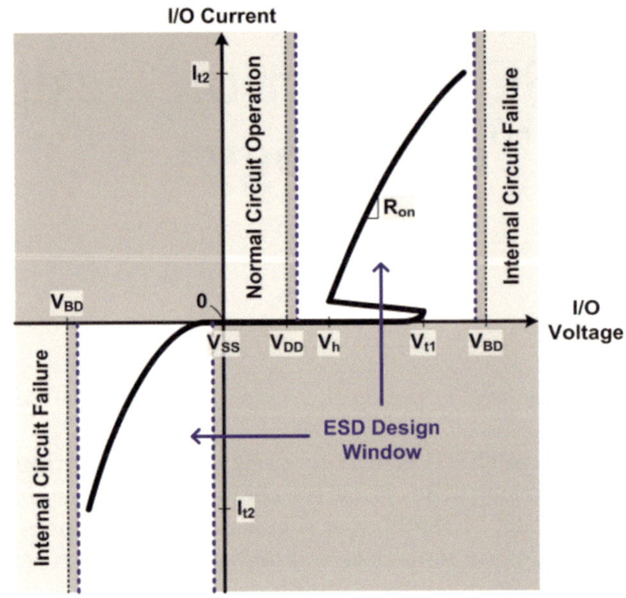

Figure 1.
ESD design window.

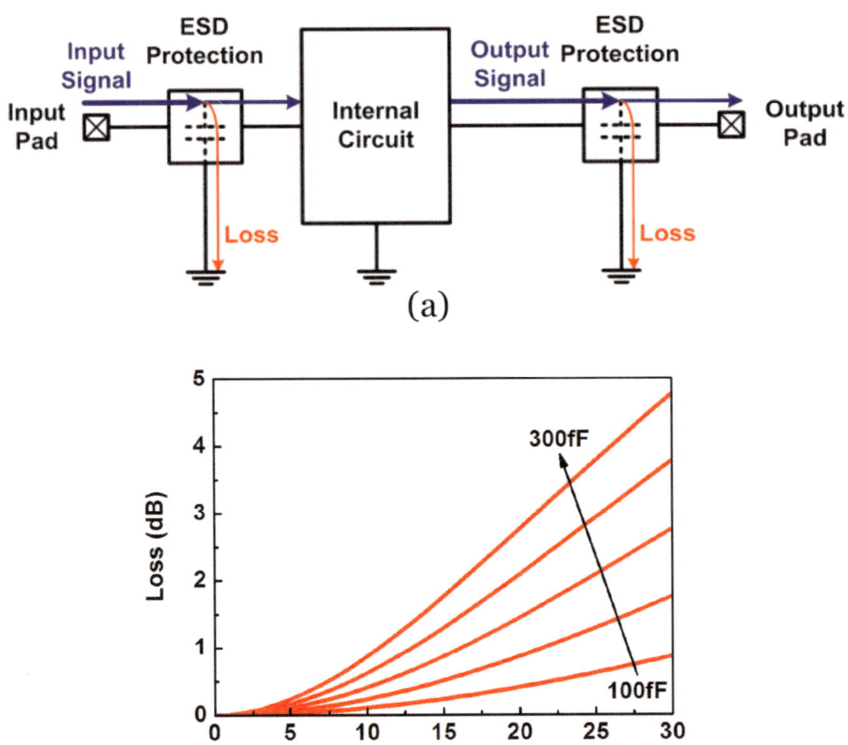

Figure 2.
(a) Parasitic capacitances seen at I/O pads cause signal loss to ground and (b) Simulated loss of parasitic capacitances.

breakdown voltage (V_{BD}) of internal circuit. First, the internal circuit normally operates between V_{DD} and V_{SS}, and the ESD protection circuit cannot turn on in this normal circuit operation region. Second, the internal circuit causes failure beyond the positive or negative V_{BD}, so the ESD protection circuit becomes invalid in this internal circuit failure region. Besides, it usually reserves some safety margin. Therefore, the ESD protection circuit must shunt ESD current with the voltage within ESD design window as shown in **Figure 1**. As ESD stresses at the I/O pad, the ESD protection circuit turns on at its trigger voltage (V_{t1}) and clamps to the holding voltage (V_h). The turn-on resistance (R_{on}) should be minimized to reduce the joule heat generated in the ESD protection circuit and enhance the current-handling ability, that is the secondary breakdown current (I_{t2}).

A typical method to enhance the current-handling ability is to widen the ESD device dimension; however, the large ESD protection device has too large parasitic capacitance to be tolerable for the high-frequency ICs. As shown in **Figure 2(a)**, the parasitic capacitances seen at the input and output (I/O) pads cause signal loss to ground. The parasitic capacitances come from not only the ESD protection circuits but also the pads and the metal connections [13, 14]. If the parasitic capacitance increases, the signal loss dramatically increases at high frequency, as shown in **Figure 2(b)**. To mitigate the performance degradation caused by the parasitic capacitance, the ESD protection circuit must carefully design. For example, a typical specification for the parasitic capacitance of input terminal of a gigahertz IC is 200fF [15].

3. ESD protection strategy

At an I/O pad of IC, it may be stressed by positive or negative ESD with grounded V_{DD} or V_{SS}. A whole-chip ESD protection design must provide the ESD current paths of all possible combinations, including the positive I/O-to-V_{DD} (PD), positive I/O-to-V_{SS} (PS), negative I/O-to-V_{DD} (ND), and negative I/O-to-V_{SS} (NS) [16]. Since the common ESD protection devices in CMOS technologies include diode, MOSFET, and silicon controlled rectifier (SCR), they are used to implement the ESD protection circuits [17]. To achieve the whole-chip ESD protection, three types of ESD protection schemes are introduced in this chapter.

Type I ESD protection circuit uses one bidirectional ESD protection device between I/O pad and V_{SS} and one bidirectional power-rail ESD clamp circuit between V_{DD} and V_{SS}, as shown in **Figure 3(a)**. The bidirectional ESD protection device could be an NMOS or SCR device. Both PS and NS ESD currents can be discharged through the ESD protection device. Besides, PD and ND ESD currents can be discharged through the ESD protection device and the power-rail ESD clamp circuit.

Type II ESD protection circuit uses two unidirectional ESD protection devices from I/O pad to V_{DD} and from V_{SS} to I/O pad, respectively, and one bidirectional power-rail ESD clamp circuit between V_{DD} and V_{SS}, as shown in **Figure 3(b)**. The unidirectional ESD protection device was a diode. Both PD and NS ESD currents can be discharged through one unidirectional ESD protection device. For the PS and ND ESD currents, they can be discharged through one ESD protection device and the power-rail ESD clamp circuit.

Type III ESD protection circuit uses a two-branched ESD protection device and an unidirectional ESD protection device between I/O pad and V_{SS} and one bidirectional power-rail ESD clamp circuit between V_{DD} and V_{SS}, as shown in **Figure 3(c)**. The two-branched ESD protection device was usually an SCR device. The PS and PD ESD currents can be discharged through the two-branched ESD protection device, and NS and ND ESD currents can be discharged through the unidirectional ESD protection device and the power-rail ESD clamp circuit.

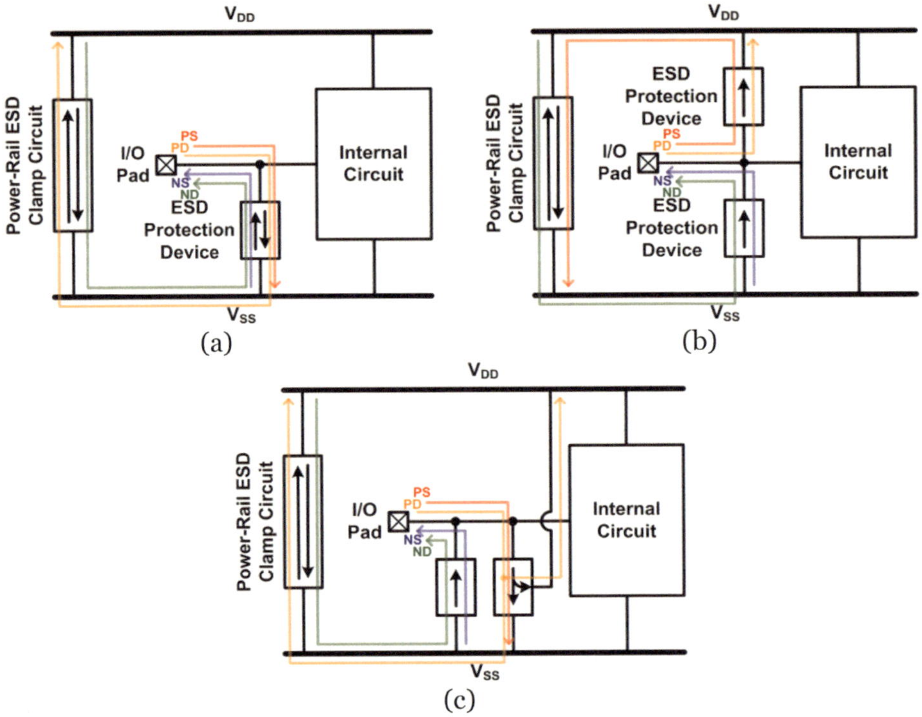

Figure 3.
ESD protection schemes: (a) type I, (b) type II, and (c) type III.

All the ESD protection devices at I/O pad should be shrunk to lower the parasitic capacitance, while the power-rail ESD clamp circuit could be as large as possible. The large-sized power-rail ESD clamp circuit can help to reduce R_{on} during ESD current discharging, but it will not cause the parasitic capacitance to the I/O pad.

4. ESD protection circuit design: Type I

A common ESD protection circuit used in CMOS technology is the grounded-gate NMOS (GGNMOS), as shown in **Figure 4(a)** [18, 19]. In this ESD protection circuit, the NMOS's gate is grounded to keep it off during normal circuit operation.

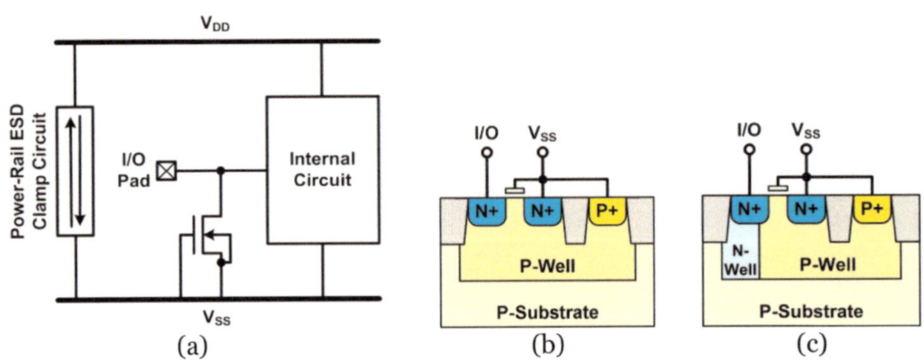

Figure 4.
(a) ESD protection circuit with GGNMOS. Device cross-sectional view of (b) GGNMOS and (c) GGNMOS with additional N-well.

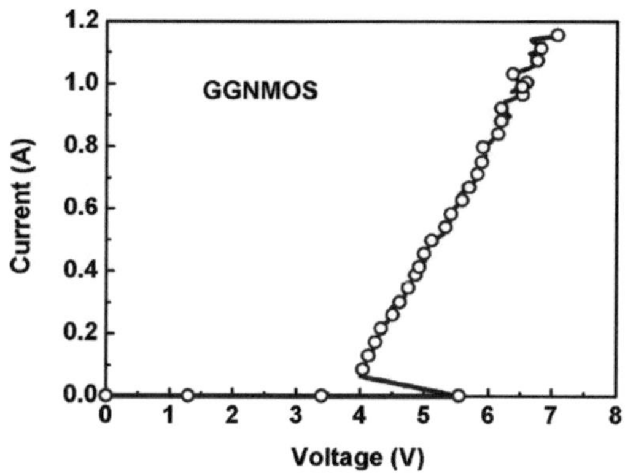

Figure 5.
TLP-measured I-V curve of a GGNMOS (W = 120 μm) in 0.18 μm CMOS technology.

The GGNMOS turns on as the positive voltage excursions above the trigger voltage (V_{t1}). **Figure 5** shows the positive I-V curve of a GGNMOS in 0.18 μm CMOS technology, which is measured by a transmission-line-pulsing (TLP) system. The TLP system with a 10 ns rise time and a 100 ns pulse width is used to investigate the turn-on behavior and the I-V characteristics in high-current regions of the test devices [20]. The trigger voltage (V_{t1}), holding voltage (V_h), and secondary breakdown current (I_{t2}) of test devices in the time domain of HBM ESD event can be extracted from the TLP-measured I-V curves. This GGMOS triggers on at 5.6 V, snapbacks to 4.0 V, and discharges ESD current until 1.1A. The GGNMOS with the help of parasitic junction diode turns on as the I/O voltage excursions below the V_{SS} voltage.

The GGNMOS is generally drawn in the multi-finger structure with central drain to save total layout area [21]. **Figure 4(b)** shows the device cross-sectional view of a single-finger GGNMOS. The multi-finger structure can be realized by combining such single-finger structures with sharing drain and source regions between every two adjacent fingers. For the high-frequency applications, the parasitic capacitance of GGNMOS has to be considered. For a given drain width (W_n) and length (L_n), the total capacitance of a GGNMOS (C_n) is given by the drain-gate overlap capacitance ($C_{overlap}$), the N+/P-well bottom junction capacitance (C_j), and the N+/P-well sidewall capacitance (C_{jsw}), according to the following equation:

$$C_n = C_{overlap} \times W_n + C_j \times W_n \times L_n + C_{jsw} \times 2 \times (W_n + L_n)$$

All the parasitic capacitance ($C_{overlap}$, C_j, and C_{jsw}) are given by the process. Besides the drain width, the L_n strongly affects the total capacitance. For high-frequency applications, the L_n needs to be optimized by reducing the contact rows, the enclosure of contacts, and the extension of silicide [22, 23]. Also the extension of silicide on drain side increases the ESD robustness of GGNMOS, it implies a larger junction area and thus induces additional parasitic capacitance of the N+/P-well bottom junction. Therefore, a trade-off between the ESD robustness and the parasitic capacitance has to be found. A possible solution to reduce the bottom capacitance with the given L_n is to use an N-well implant below the N+ drain, as shown in **Figure 4(c)**. Most of the bottom N+/P-well capacitance is then replaced by an N-well/P-well sidewall capacitance and N-well/P-substrate bottom capacitance.

Instead of GGNMOS, gate-coupled NMOS and substrate-coupled NMOS have also been used as ESD protection circuit [24]. However, the parasitic capacitance of

MOS-based ESD protection device is usually too large to be tolerable for the high-frequency circuits.

An alternative ESD protection device used in Type I ESD protection circuit is a silicon controlled rectifier (SCR) [25]. The SCR device has been reported to be useful for ESD protection in high-frequency circuits due to its higher ESD robustness within a smaller layout area and lower parasitic capacitance [22]. Besides, the SCR device can be safely used without latchup danger in advanced CMOS technologies with low supply voltage [26]. The equivalent circuit of the SCR consists of a PNP BJT and an NPN BJT, as shown in **Figure 6(a)**. As ESD zapping from I/O to V_{SS}, the positive-feedback regenerative mechanism of PNP and NPN results in the SCR device highly conductive to make SCR very robust against ESD stresses. The device structure of the SCR device is illustrated in **Figure 6(b)**. The I/O pad is connected to the first P+ and the pickup N+, which is formed in the N-well. The V_{SS} pad is connected to the second N+ and the pickup P+, which are formed in the nearby P-well. The SCR path between I/O and V_{SS} consists of P+, N-well, P-well, and N+. Besides, the parasitic diode path from V_{SS} to I/O consists of P-well and N-well. The SCR with the help of P-well/N-well junction diode turns on as the I/O voltage excursions below the V_{SS} voltage.

Figure 7 shows the TLP-measured positive I-V curve of an SCR in 0.18 μm CMOS technology. This SCR triggers on at 16.7 V, snapbacks to 2.1 V, and discharges ESD current until 9.5A. The main drawback of SCR device is the higher trigger voltage and thus the slower turn-on speed. Research works have demonstrated that separation of the N-well and P-well junction can play an important role. The typical SCR device uses the shallow trench isolation (STI) to separate the N-well and P-well. To reduce the trigger voltage of an SCR device, a gate-bounded SCR has been reported, as shown in **Figure 6(c)** [27].

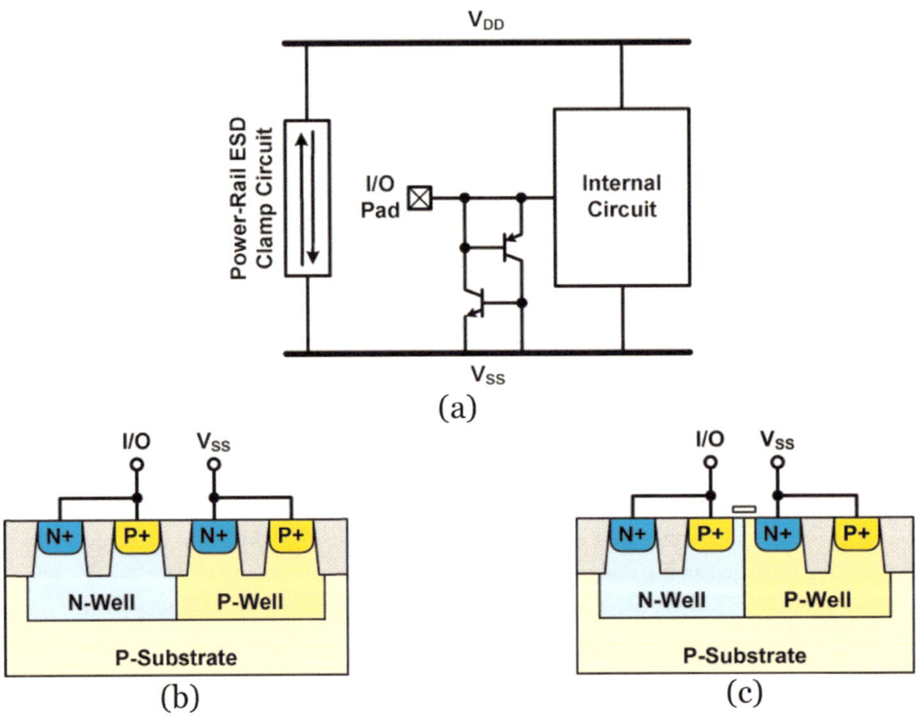

Figure 6.
(a) ESD protection circuit with SCR. Device cross-sectional view of (b) STI-bounded SCR and (c) gate-bounded SCR.

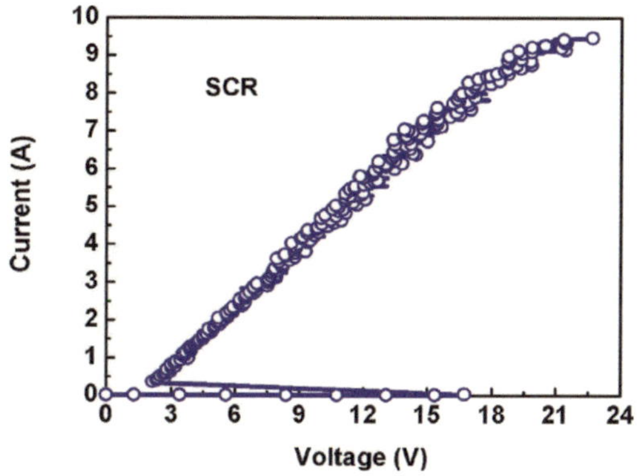

Figure 7.
TLP-measured I-V curve of an SCR (W = 120 μm) in 0.18 μm CMOS technology.

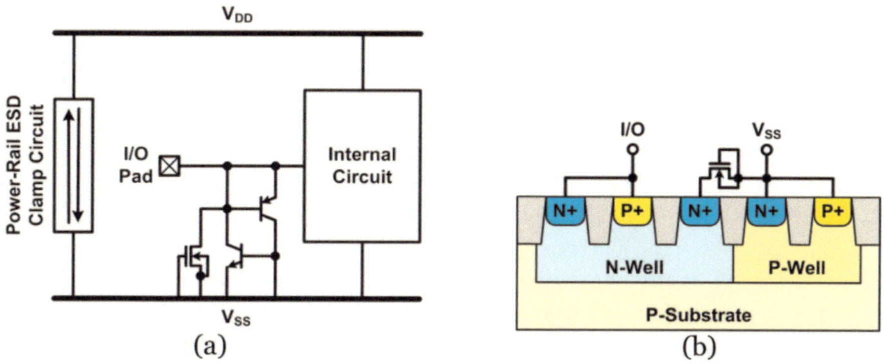

Figure 8.
(a) ESD protection circuit with GGNMOS-triggered SCR and (b) device cross-sectional view of GGNMOS-triggered SCR.

Another alternative method to reduce the trigger voltage of an SCR device uses the substrate-triggered technique. The trigger signal can be sent into the base terminal of PNP or NPN to enhance the turn-on speed. Some circuit design techniques are reported to enhance the turn-on efficiency of SCR devices, such as the gate-coupled, substrate-triggered, diode-triggered, and gate-grounded-NMOS-triggered (GGNMOS-triggered) techniques [28–30]. **Figure 8(a)** shows the schematic of a GGNMOS-triggered SCR device, and **Figure 8(b)** shows its device cross-sectional view. The GGNMOS is connected between the second N+ in the N-well and V_{SS}. The trigger current is drawn from the N-well (base of PNP) to V_{SS} through the GGNMOS. Similarly, the trigger device can be connected between I/O pad and the base and NPN, but the trigger device will also add the parasitic capacitance to I/O. A diode string could also be used as the trigger device, and its parasitic capacitance is lower than the GGNMOS.

Recently, an inductor-assisted diode-triggered SCR (LASCR) has been presented to further reduce the parasitic capacitance [31]. As shown in **Figure 9**, the LASCR consists of an SCR, an inductor, and a diode string. The ESD current path from I/O to V_{SS} consists of P+/N-well/P-well/N+ SCR. The diode string drawn the trigger current from the N-well (base of PNP) to V_{SS} is used to enhance the turn-on efficiency of SCR. As the I/O voltage excursions below the V_{SS} voltage, the ESD current path consists of P-well/N-well diode and inductor.

Under normal circuit operating condition, the inductor can resonate with the parasitic capacitance, and hence the signal loss can be compensated.

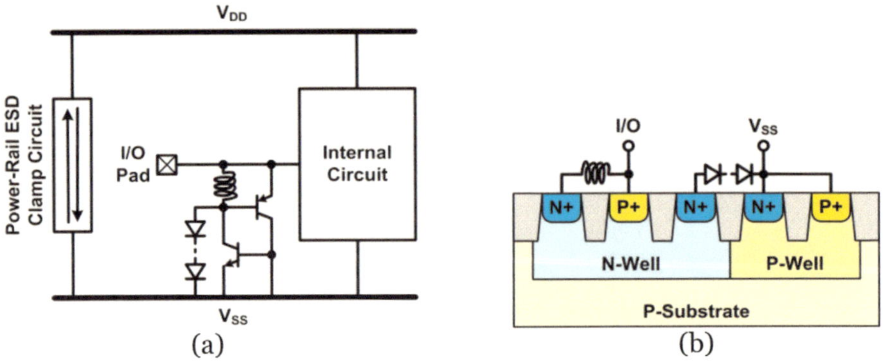

Figure 9.
(a) ESD protection circuit with LASCR and (b) device cross-sectional view of LASCR.

Once the dimension of SCR has been chosen, the inductance (L) can be designed to minimize the high-frequency performance degradation by using the following equation:

$$L = \frac{1}{C_{P+/N-well} \times (2\pi f_o)^2}$$

where $C_{P+/N\text{-}well}$ is the parasitic capacitance of P+/N-well junction, and f_o is the operating frequency. For example, the dimension of SCR is selected to be 30 μm, and the $C_{P+/N\text{-}well}$ in a 0.18 μm CMOS process is ~60fF around 30GHz. Therefore, the required L for 30GHz applications is 460pH.

Figure 10(a) shows the TLP-measured I-V curves of LASCR with 3 and 5 diodes in diode string (LASCR_3D and LASCR_5D) in a 0.18 μm CMOS process. The LASCR_3D triggers on at 5.2 V, snapbacks to 2.9 V, and discharges ESD current until 2.4A, while LASCR_5D triggers on at 7.6 V, snapbacks to 2.9 V, and discharges ESD current until 2.1A. The trigger voltage can be adjusted by adding or reducing the diode numbers. The holding voltage of both LASCR devices exceed V_{DD} (1.8 V in the given CMOS process), which is safe from latchup event.

The signal losses of both LASCR devices are measured through the on-wafer two-port measurement. The measured loss versus frequencies of both LASCR

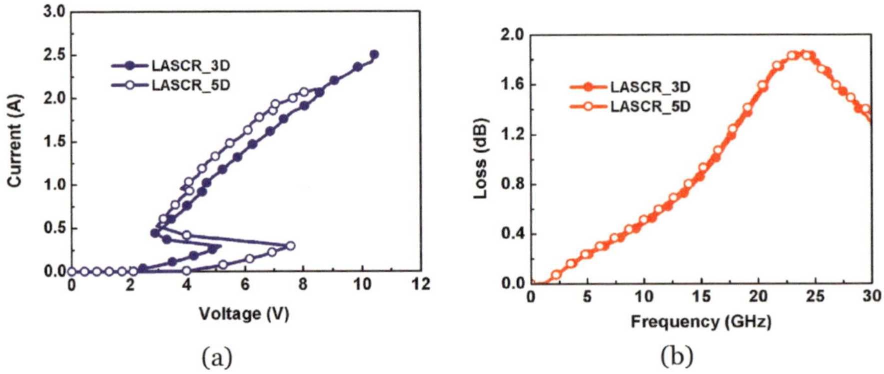

Figure 10.
(a) TLP-measured I-V curves and (b) loss of LASCR (W = 30 μm) with 3 and 5 trigger diodes in 0.18 μm CMOS technology.

devices is shown in **Figure 10(b)**. The LASCR devices exhibit sufficiently low loss even if the frequency is up to 30GHz. Therefore, LASCR can be a good solution for ESD protection of high-speed applications.

5. ESD protection circuit design: Type II

Diode is a typical ESD protection device with unidirectional discharging path [32, 33]. A dual-diode ESD protection circuit for high-frequency applications is shown in **Figure 11(a)**, where two ESD diodes at I/O pad are cooperated with the turn-on efficient power-rail ESD clamp circuit to discharge ESD current in the forward-biased condition [13, 34].

In the CMOS process, the choice for ESD protection diodes includes P+/N-well, N+/P-well, and N-well/P-well diodes. The P+/N-well diode, as shown in **Figure 11(b)**, is used between I/O pad and V_{DD}. For the N-well/P-well diode, it may occupy larger layout area than the N+/P-well diode. Thus, the N+/P-well diode, as shown in **Figure 11(c)**, is used between V_{SS} and I/O pad.

The typical diodes use the STI to separate the PN junctions. Besides the STI-bounded diodes, the gate-bounded diodes have been reported, as shown in **Figure 11(d)** and **(e)**. The gate-bounded diodes were introduced by Voldman in order to improve the ESD robustness of STI bounded diodes [35].

In order to reduce the parasitic capacitance or provide the large signal-swing tolerance, the ESD protection diodes in stacked configuration have been presented [36, 37], as shown in **Figure 12(a)**. The device cross-sectional views of the conventional stacked diodes are shown in **Figure 12(b)** and **(c)**. Two P+/N-well diodes (stacked P diodes) can apply to I/O-to-V_{DD}, and two N+/P-well diodes (stacked N diodes) can apply to V_{SS}-to-I/O, as shown in **Figure 12(b)** and **(c)**, respectively. With the stacked diodes, the junction capacitances are connected in series, and the

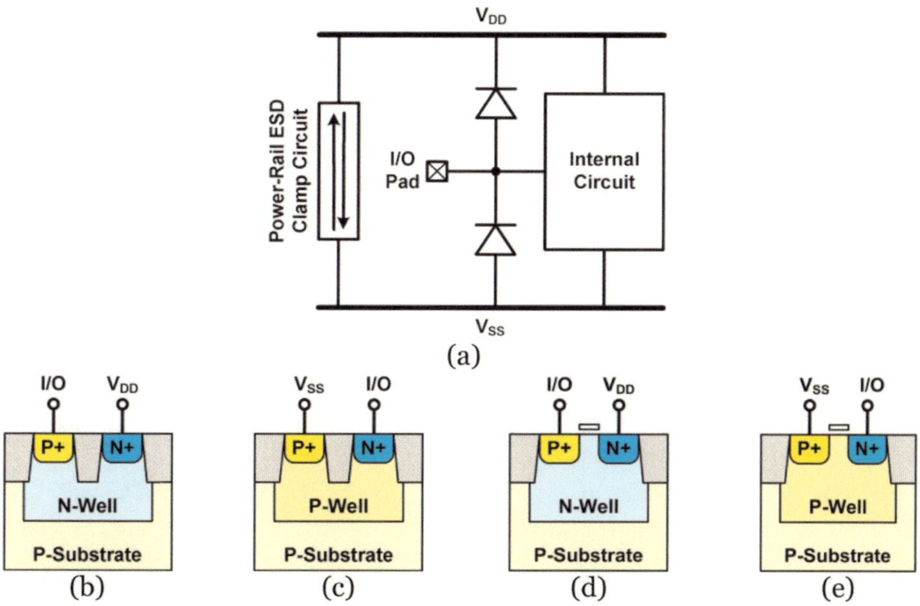

Figure 11.
(a) ESD protection circuit with diodes. Device cross-sectional view of (b) STI-bounded P+/N-well diode, (c) STI-bounded N+/P-well diode, (d) gate-bounded P+/N-well diode, and (e) gate-bounded N+/P-well diode.

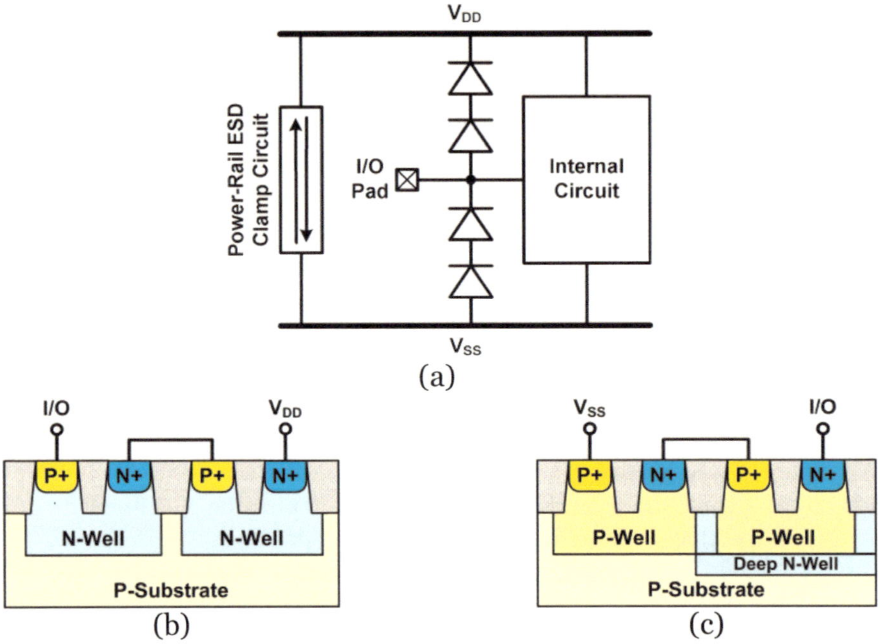

Figure 12.
ESD protection circuit with stacked diodes. (a) ESD protection circuit with stacked diodes. Device cross-sectional view of (b) stacked P+/N-well diode and (c) stacked N+/P-well diode.

overall parasitic capacitance becomes smaller. However, the stacked configuration is adverse to ESD protection because the overall turn-on resistance and the clamping voltage of the stacked diodes during ESD stresses are increased as well.

For effective ESD protection, the stacked diodes with embedded SCR (SDSCR) have been presented [38, 39]. The SCR device has been reported to be useful for ESD protection with low turn-on resistance, low parasitic effects, and high ESD robustness. The stacked diodes with embedded SCR are illustrated in **Figure 13**. In this design, a P+/N-well diode and an N+/P-well diode are stacked, and a P+/N-well/P-well/N+ SCR is embedded to form the ESD current path. A deep N-well structure is used to isolate the P-well region from the common P-substrate, so the SDSCR can apply to I/O-to-V_{DD} or V_{SS}-to-I/O. In the beginning of ESD stress, the initial current will be discharged through the stacked diodes, and then the primary current will be discharged through the embedded SCR. The stacked diodes also play the role of trigger circuit of SCR, because the current drawn from N-well and

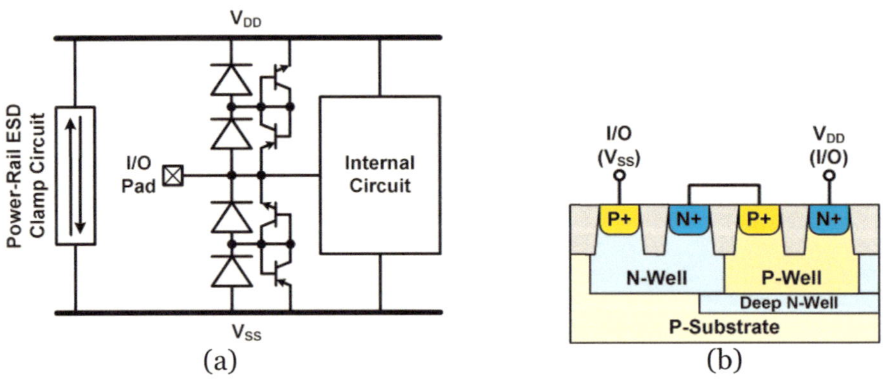

Figure 13.
(a) ESD protection circuit with SDSCR and (b) device cross-sectional view of SDSCR.

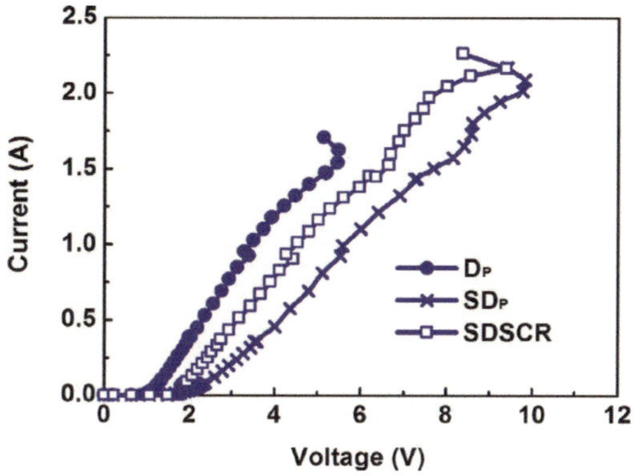

Figure 14.
TLP-measured I-V curves of D_P, SD_P, and SDSCR (W = 20 μm) in 0.18 μm CMOS technology.

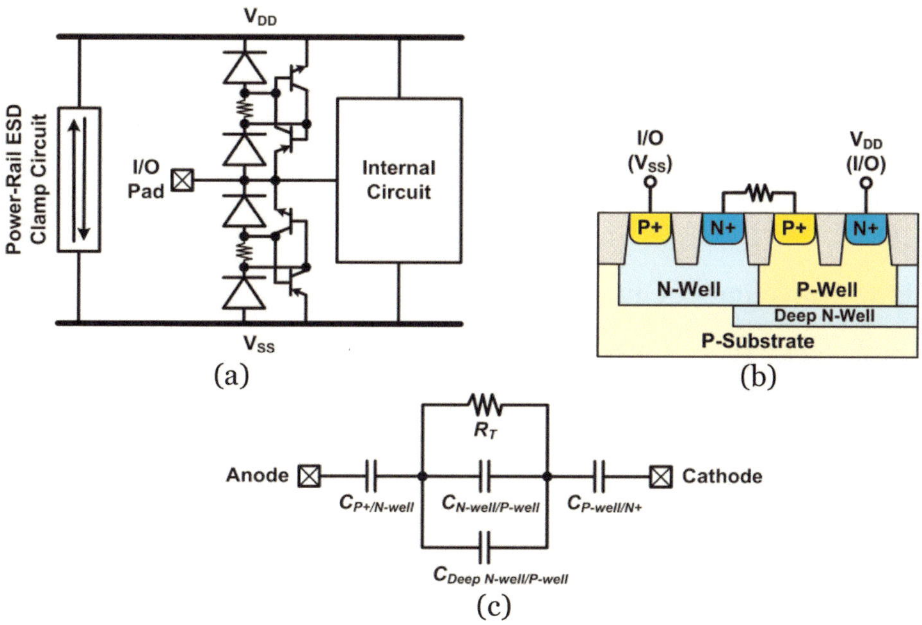

Figure 15.
(a) ESD protection circuit with RTSCR. (b) device cross-sectional view and (c) simplified model of RTSCR.

injected into P-well can also trigger the PNP and the NPN of SCR. **Figure 14** shows the TLP-measured I-V curves of P+/N-well diode (D_P), stacked P+/N-well diodes (SD_P), and stacked diodes with embedded SCR (SDSCR) in a 0.18 μm CMOS process. We can find that turn-on resistance or the clamping voltage of single diode is much lower than that of the stacked diodes. The embedded SCR can help to slightly reduce the turn-on resistance and the clamping voltage of the stacked diodes. In fact, some layout skills can be used to further improve the turn-on efficient of the stacked diodes with embedded SCR [40].

Recently, a similar structure of the stacked diodes with embedded SCR, where a resistor uses to separate two diodes, has been reported [41]. The resistor acts as the trigger element of SCR, so the device is named resistor-triggered SCR (RTSCR). **Figure 15(a)** and **(b)** shows the schematic and the device cross-sectional view of RTSCR. The resistor can also reduce the parasitic capacitance of the ESD protection circuit. Considering the simplified SCR model by using junction capacitances, as shown in **Figure 15(c)**, the equivalent capacitance seen at anode or cathode of RTSCR can be calculated by the following equation:

$$C_{RTSCR} = \frac{\mathrm{Im}(Y_{RTSCR})}{\omega} = \frac{\mathrm{Im}\left(\frac{1}{\frac{1}{j\omega C_{P+/N-Well}} + \frac{1}{\frac{1}{R_T} + j\omega C_{P-Well/N-Well(Deep\ N-Well)}} + \frac{1}{j\omega C_{P-Well/N+}}}\right)}{\omega}$$

where Y_{RTSCR} denotes the admittance of the RTSCR, R_T is the resistance, and $C_{P+/N\text{-}Well}$, $C_{P\text{-}Well/N\text{-}Well(Deep\ N\text{-}Well)}$ and $C_{P\text{-}Well/N+}$ denote the junction capacitances. To simplify the above equation, the junction capacitance is rewritten to C_J, and then the parasitic capacitance of the RTSCR can be expressed by the following equation:

$$C_{RTSCR} = \mathrm{Im}\left(\frac{1}{\frac{2}{jC_J} + \frac{1}{\frac{1}{\omega R_T} + jC_J}}\right) = \frac{\frac{2}{C_J} + \frac{\omega^2 R_T{}^2 C_J}{1+\omega^2 R_T{}^2 C_J{}^2}}{\left(\frac{2}{C_J} + \frac{\omega^2 R_T{}^2 C_J}{1+\omega^2 R_T{}^2 C_J{}^2}\right)^2 + \left(\frac{\omega R_T}{1+\omega^2 R_T{}^2 C_J{}^2}\right)^2} \approx \frac{C_J}{2 + \frac{3}{2}\omega^2 R_T{}^2 C_J{}^2}$$

It can be noted that the parasitic capacitance of the RTSCR can be reduced by adding the resistor. Generally, the capacitance reduction of RTSCR can be up to 30%. Therefore, the ESD protection circuit with dual RTSCRs can be used for high-frequency applications.

6. ESD protection circuit design: Type III

Figure 16(a) shows another SCR-based ESD protection circuit [13]. The typical SCR device in CMOS process consists of P+, N-well, P-well, and N+. Instead of connecting the N-well to I/O pad, connecting the N-well to V_{DD} avoids the parasitic capacitance or noise coupling from P-substrate or P-well to N-well and I/O [42]. As shown in **Figure 16(b)**, the I/O pad is connected to the first P+, which is formed in the N-well. The pickup N+ in the N-well is biased to V_{DD}. The V_{SS} pad is connected to the second N+ and the pickup P+, which are formed in the nearby P-well. The SCR path between I/O and V_{SS} consists of P+, N-well, P-well, and N+. Besides, the parasitic diode path from I/O to V_{DD} consists of P+ and N-well. In this structure, the PS and the PD ESD currents can be discharged through the SCR path and its parasitic diode path. The NS and the ND ESD currents need reverse diode and power-rail ESD clamp circuit to form their discharging paths.

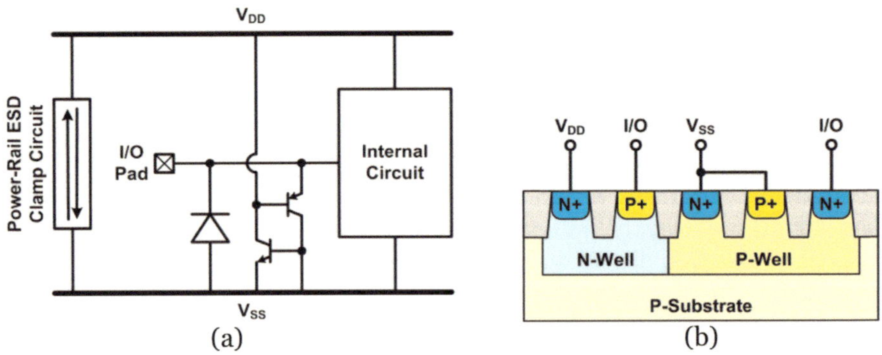

Figure 16.
(a) ESD protection circuit with SCR and diode and (b) device cross-sectional view of SCR and diode.

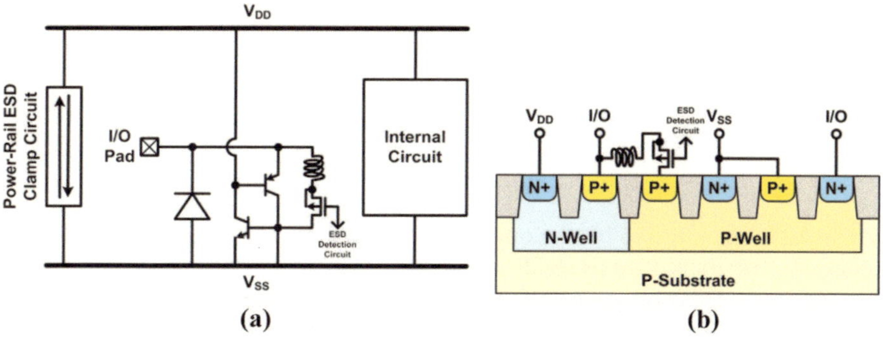

Figure 17.
ESD protection circuit with LTSCR and reverse diode. (a) ESD protection circuit with LTSCR and reverse diode and (b) device cross-sectional view of LTSCR and reverse diode.

The SCR device in this ESD protection circuit still has the drawbacks of higher trigger voltage and the slower turn-on speed. The circuit design techniques, including the gate-coupled, substrate-triggered, diode-triggered, and GGNMOS-triggered techniques can be used to enhance the turn-on efficiency of SCR device. Of course, the capacitive triggering device increases the total parasitic capacitance seen at the I/O pad, even if the triggering device is not directly connected to I/O. Recently, an SCR device with inductive triggering device has been presented [43]. That inductor-triggered SCR (LTSCR) is proposed for ESD protection of high-frequency applications to achieve low high-frequency performance degradation, low trigger voltage, and high ESD robustness. In this design, the inductor provides a current path to trigger the SCR device, and it can also compensate the parasitic capacitance of ESD protection devices.

Figure 17(a) shows the ESD protection circuit with an LTSCR and a reverse diode. This design consists of an SCR device and a reverse diode as the main ESD current path, and an inductor (L_{trig}), a MOS transistor (M_{trig}), and an RC-based ESD detection circuit as the trigger circuit. The initial-on PMOS transistor is selected for M_{trig} to quickly pass the trigger current to SCR device [44]. The positive and negative ESD current discharging paths for the I/O pad are provided by the SCR and the reverse diode. **Figure 17(b)** shows the device cross-sectional view of inductor-triggered SCR. Under ESD stress conditions, the inductor and PMOS are used to provide the trigger path between the I/O pad and the base of NPN of the SCR device. When the trigger current is sent into the base of NPN of the SCR device, the SCR device can be quickly triggered on to discharge the ESD current from the I/O pad to V_{SS}. The ESD detection circuit usually uses RC timer to distinguish the ESD-stress conditions from the normal circuit operating conditions, and the PMOS transistor is well controlled to turn on or off by the ESD detection circuit. Under normal circuit operating conditions, the inductor can compensate the parasitic capacitance of SCR and diode.

In this circuit, the dimensions of the inductor (L_{trig}), PMOS transistor (M_{trig}), SCR device, and reverse diode can be designed to minimize the high-frequency performance degradation. Since the capacitor used in power-rail ESD clamp circuit is large enough to keep the node between R and C at AC ground under normal circuit operating conditions, the impedance of the trigger path (Z_{trig}) seen at the I/O pad to ground can be calculated as:

$$Z_{\text{trig}} \approx j\omega L_{trig} + \frac{1}{j\omega C_{trig}} = j\omega\left(L_{trig} - \frac{1}{\omega^2 C_{trig}}\right)$$

where C_{trig} is the sum of gate-to-source, gate-to-body, and drain-to-body capacitances of the PMOS. The resonance angular frequency (ω_o) can be obtained by

$$\omega_o = \frac{1}{\sqrt{\left(L_{trig} - \frac{1}{\omega_o{}^2 C_{trig}}\right) C_{ESD}}}$$

where ω_o is designed to be the operating frequency, and C_{ESD} is the parasitic capacitance contributed by the SCR and diode. The sizes of SCR and diode depend on the required ESD robustness, while the size of M_{trig} transistor depends on the required trigger current. Once the sizes of M_{trig} transistor, SCR, and diode have been chosen, the required inductance (L_{trig}) can be determined.

7. Conclusion

A comprehensive review in the field of ESD protection design for high-frequency integrated circuits is presented in this chapter. Besides improving the ESD robustness, the parasitic effects from ESD protection devices must be minimized or canceled to optimize the high-frequency performance simultaneously. Furthermore, the ESD protection circuits and high-frequency circuits can be co-designed to achieve both good circuit performance and high ESD robustness. The on-chip ESD protection designs for high-frequency circuits will be continuously an important design task in CMOS technology.

Author details

Chun-Yu Lin
Department of Electrical Engineering, National Taiwan Normal University, Taipei, Taiwan

*Address all correspondence to: cy.lin@ieee.org

References

[1] Rangan S, Rappaport T, Erkip E. Millimeter-wave cellular wireless networks: Potentials and challenges. Proceedings of the IEEE. 2014;**102**: 366-385

[2] Fritsche D, Tretter G, Carta C, Ellinger F. Millimeter-wave low-noise amplifier design in 28-nm low-power digital CMOS. IEEE Transactions on Microwave Theory and Techniques. 2015;**63**:1910-1922

[3] Abidi A. CMOS microwave and millimeter-wave ICs: The historical background. In: Proceedings of the IEEE International Symposium on Radio-Frequency Integration Technology (RFIT). 2014

[4] Gossner H. Design for ESD protection at its limits. In: Proceedings of the Symposium on VLSI Technology. 2013

[5] Li J, Mishra R, Shrivastava M, Yang Y, Gauthier R, Russ C. Technology scaling effects on the ESD performance of silicide-blocked PMOSFET devices in nanometer bulk CMOS technologies. In: Proceedings of the Electrical Overstress/ Electrostatic Discharge Symposium (EOS/ESD). 2011

[6] Wang A, Feng H, Zhan R, Xie H, Chen G, Wu Q, et al. A review on RF ESD protection design. IEEE Transactions on Electron Devices. 2005;**52**:1304-1311

[7] Natarajan M, Linten D, Thijs S, Jansen P, Trémouilles D, Jeamsaksiri W, Nakaie T, Sawada M, Hasebe T, Decoutere S, Groeseneken G. RFCMOS ESD protection and reliability. In: Proceedings of the International Symposium on the Physical and Failure Analysis of Integrated Circuits (IPFA); 2005. p. 59-66

[8] Voldman S. ESD: RF Technology and Circuits. Chichester: John Wiley & Sons; 2006

[9] Ker M, Lin C, Hsiao Y. Overview on ESD protection designs of low-parasitic capacitance for RF ICs in CMOS technologies. IEEE Transactions on Device and Materials Reliability. 2011; **11**:207-218

[10] Voldman S. ESD: Analog Circuits and Design. Chichester: John Wiley & Sons; 2015

[11] Amerasekera A, Duvvury C. ESD in Silicon Integrated Circuits. Chichester: John Wiley & Sons; 2002

[12] Li J, Chatty K, Gauthier R, Mishra R, Russ C. Technology scaling of advanced bulk CMOS on-chip ESD protection down to the 32nm node. In: Proceedings of the Electrical Overstress/Electrostatic Discharge Symposium (EOS/ESD). 2009

[13] Galy P, Jimenez J, Meuris P, Schoenmaker W, Dupuis O. ESD RF protections in advanced CMOS technologies and its parasitic capacitance evaluation. In: Proceedings of the IEEE International Conference on IC Design & Technology (ICICDT); 2011

[14] Peng B, Lin C. Low-loss I/O pad with ESD protection for K/Ka-bands applications in nanoscale CMOS process. IEEE Transactions on Circuits and Systems II: Express Briefs. 2018;**65**: 1475-1479

[15] Soldner W, Streibl M, Hodel U, Tiebout M, Gossner H, Schmitt-Landsiedel D, et al. RF ESD protection strategies: Codesign vs. low-C protection. In: Proceedings of the Electrical Overstress/Electrostatic Discharge Symposium (EOS/ESD). 2005

[16] Ker M. Whole-chip ESD protection design with efficient VDD-to-VSS ESD clamp circuit for submicron CMOS

VLSI. IEEE Transactions on Electron Devices. 1999;**46**:173-183

[17] Voldman S. ESD: Circuits and Devices. Chichester: John Wiley & Sons; 2015

[18] Wang W, Dong S, Zhong L, Zeng J, Yu Z, Liu Z. GGNMOS as ESD protection in different nanometer CMOS process. In: Proceedings of the IEEE International Conference on Electron Devices and Solid-State Circuits (EDSSC); 2014

[19] Paul M, Russ C, Kumar B, Gossner H, Shrivastava M. Physics of current filamentation in ggNMOS devices under ESD condition revisited. IEEE Transactions on Electron Devices. 2018; **65**:2981-2989

[20] Maloney T, Khurana N. Transmission line pulsing techniques for circuit modeling of phenomena. In: Proceedings of the Electrical Overstress/ Electrostatic Discharge Symposium (EOS/ESD). 1985

[21] Lee J, Wu K, Huang S, Tang C. The dynamic current distribution of a multi-fingered GGNMOS under high current stress and HBM ESD events. In: Proceedings of the IEEE International Reliability Physics Symposium (IRPS); 2006

[22] Richier C, Salome P, Mabboux G, Zaza I, Juge A, Mortini P. Investigation on different ESD protection strategies devoted to 3.3 V RF applications (2 GHz) in a 0.18 μm CMOS process. Journal of Electrostatics. 2002;**54**:55-71

[23] Chen T, Ker M, Wu C. Experimental investigation on the HBM ESD characteristics of CMOS devices in a 0.35-μm silicided process. In: Proceedings of the IEEE International Symposium on VLSI Technology, Systems, and Applications. VLSI-TSA; 1999; Hsinchu

[24] Song B, Han Y, Li M, Liou J, Dong S, Guo W, Huang D, Ma F, Miao M. Design analysis of novel substrate-triggered GGNMOS in 65nm CMOS process. In: Proceedings of the IEEE International Symposium on the Physical and Failure Analysis of Integrated Circuits (IPFA); 2010

[25] Ker M, Hsu K. Overview of on-chip electrostatic discharge protection design with SCR-based devices in CMOS integrated circuits. IEEE Transactions on Device and Materials Reliability. 2005;**5**:235-249

[26] Dai C, Chen S, Linten D, Scholz M, Hellings G, Boschke R, Karp J, Hart M, Groeseneken G, Ker M, Mocuta A, Horiguchi N. Latchup in bulk FinFET technology. In: Proceedings of the IEEE International Reliability Physics Symposium (IRPS); 2017

[27] Chang T, Hsu Y, Tsai T, Tseng J, Lee J, Song M. High-k metal gate-bounded silicon controlled rectifier for ESD protection. In: Proceedings of the Electrical Overstress/Electrostatic Discharge Symposium (EOS/ESD). 2012

[28] Ker M, Hsu K. Substrate-triggered SCR device for on-chip ESD protection in fully silicided sub-0.25-μm CMOS process. IEEE Transactions on Electron Devices. 2003;**50**:397-405

[29] Jang S, Gau M, Lin J. Novel diode-chain triggering SCR circuits for ESD protection. Solid-State Electronics. 2000;**44**:1297-1303

[30] Russ C, Mergens M, Verhaege K, Armer J, Jozwiak P, Kolluri G, et al. GGSCRs: GGNMOS triggered silicon controlled rectifiers for ESD protection in deep sub-micron CMOS processes. In: Proceedings of the Electrical Overstress/ Electrostatic Discharge Symposium (EOS/ESD). 2001; Portland

[31] Lin C, Chang R. Design of ESD protection device for K/Ka-band applications in nanoscale CMOS process. IEEE Transactions on Electron Devices. 2015;**62**:2824-2829

[32] Bhatia K, Jack N, Rosenbaum E. Layout optimization of ESD protection diodes for high-frequency I/Os. IEEE Transactions on Device and Materials Reliability. 2009;**9**:465-475

[33] Chen S, Linten D, Lee J, Scholz M, Hellings G, Sibaja-Hernandez A, Boschke R, Song M, See Y, Groeseneken G, Thean A. Proceedings of the IEEE International Electron Devices Meeting (IEDM); 2014

[34] Yeh C, Ker M, Liang Y. Optimization on layout style of ESD protection diode for radio-frequency front-end and high-speed I/O interface circuits. IEEE Transactions on Device and Materials Reliability. 2010;**10**: 238-246

[35] Voldmm S, Schulz R, Howard J, Gross V, Wu S, Yapsir A, et al. CMOS-on-SOI ESD protection networks. In: Proceedings of the Electrical Overstress/ Electrostatic Discharge Symposium (EOS/ESD). 1996

[36] Ruberto M, Degani O, Wail S, Tendler A, Fridman A, Goltman G. A reliability-aware RF power amplifier design for CMOS radio chip integration. In: Proceedings of the IEEE International Reliability Physics Symposium (IRPS); 2008

[37] Son M, Park C. Electrostatic discharge protection devices with series connection using distributed cell-based diodes. Electronics Letters. 2014;**50**:168-170

[38] Lin C, Fan M, Ker M, Chu L, Tseng J, Song M. Improving ESD robustness of stacked diodes with embedded SCR for RF applications in 65-nm CMOS. In: Proceedings of the IEEE International Reliability Physics Symposium (IRPS); 2014

[39] Lin C, Fu W. Diode string with reduced clamping voltage for efficient on-chip ESD protection. IEEE Transactions on Device and Materials Reliability. 2016;**16**:688-690

[40] Lin C, Fan M. Optimization on layout style of diode stackup for on-chip ESD protection. IEEE Transactions on Device and Materials Reliability. 2014; **14**:775-777

[41] Lin C, Chen C. Resistor-triggered SCR device for ESD protection in high-speed I/O interface circuits. IEEE Electron Device Letters. 2017;**38**:712-715

[42] Afzali-Kusha A, Nagata M, Verghese N, Allstot D. Substrate noise coupling in SoC design: Modeling, avoidance, and validation. Proceedings of the IEEE. 2006;**94**:2109-2138

[43] Lin C, Chu L, Ker M. ESD protection design for 60-GHz LNA with inductor-triggered SCR in 65-nm CMOS process. IEEE Transactions on Microwave Theory and Techniques. 2012;**60**:714-723

[44] Ker M, Chen S. Implementation of initial-on ESD protection concept with PMOS-triggered SCR devices in deep-submicron CMOS technology. IEEE Journal of Solid-State Circuits. 2007;**42**: 1158-1168

Section 2

Breakdown in Micro-gaps

Chapter 3

Electrical Breakdown Behaviors in Microgaps

Guodong Meng and Yonghong Cheng

Abstract

The study of electrical breakdown behaviors in microgaps has drawn intensive attention around the world due to the miniaturization of electronic devices that allows electronic circuits to be packaged more densely, making possible compact computers, advanced radar and navigation systems, and other devices that use very large numbers of components. Therefore, a clear understanding of the electrical breakdown behaviors in microgaps is required to avoid the dielectric breakdown or to trigger the breakdown at microscale. This chapter introduces the significance of understanding breakdown characterization and reliability assessment for electrostatically actuated devices, magnetic recording devices, photomasks, RF MEMS switches, and micromachines and points out the derivation of the classical Paschen's law at microscale. Then it summarizes the state-of-the-art research work on the methodology, influencing factors, dynamics, and physical mechanisms of electrical breakdown in microgaps, which is expected to expand the general knowledge of electrical breakdown to the microscale regime or more and benefits the reliability assessment and ESD protection of microscale and nanoscale devices.

Keywords: electrical breakdown behaviors, microscale, Townsend avalanche, field emission, influencing factors, dynamics, physical mechanism

1. Introduction

1.1 Background and motivation

Device miniaturization has revolutionized electronics, allowing denser packaging of electronic circuits to make possible compact computers, advanced radar and navigation systems, and other devices that use very large numbers of components [1]. In practical applications, micro-electromechanical systems (MEMSs), like micromachines and micro-mirror arrays, function by electrostatic actuation [2, 3], while the electronic devices, like photomasks [4, 5] and magnetoresistive (MR), giant magnetoresistive (GMR), and tunneling magnetoresistive (TMR) devices used in the magnetic recording industry [6–8], are at risk of accumulating static charges and the consequent threats of electrostatic discharge (ESD); both the microdevices and microstructures are associated with a strong electric field strength within microgaps [9]. For instance, the high operating voltages required for RF MEMS switches [10–13], micro-motors [14, 15] and micro-mirror [16, 17] can create sparking or breakdown across microgap structures due to electrical overstress (EOS) that may damage or destroy sensitive equipment, especially when the devices are subjected to a complex electromagnetic environment [7, 18].

Besides, the photomasks, which are used in front-end semiconductor photolithography processing to project a desired pattern onto the wafer surface, could become charged and a spark can occur either due to the real charge on the chrome guard ring or the induced charge caused by fields from surface charge on the quartz [4, 19]. Meanwhile, multiple applications in combustion, chemistry, biology, and medicine require the intentional creation of microplasmas or microdischarges [20, 21]. For instance, various microelectric propulsion systems have been proposed for ultra-small satellites, including Hall thrusters or pulsed plasma thrusters [22–25], which utilize microdischarges. As the devices are getting smaller from microscale to nanoscale and even molecular scale, the reliability assessment and underlying physics about the static charge and ESD events draw increasing attentions from both academics and industry [26–28]. Hence, predicting dielectric breakdown thresholds and figuring out the physical mechanism of microgap structures are critical to avoid undesired discharge or improve the microplasma performance, which would be of great interest to the microelectronic and plasma communities.

1.2 Derivation from the classical Paschen's law

The gas breakdown phenomenon was recognized ever since the creation of human beings thousands of years ago, but firstly systematically investigated by German physicist Paschen in 1889 [29]. Through conducting a series of electrical discharge experiments, Paschen established the widely used Paschen's law, which described the relationship between the breakdown voltage V_{bd} and the product of the pressure p and gap length d. Since then, Paschen's law has been employed for predicting breakdown thresholds and insulation performance of power equipment, electronic devices, etc.

Generally, Paschen's law could be explained by the Townsend avalanche mechanism, which considers that the electrons collide and ionize with neutral particles (α process) and positive ions bombard the cathode and generate secondary electrons (γ process), which are the primary processes during the discharge. Paschen's law could be described by the equation

$$U_{bd} = \frac{Bpd}{\ln(Apd) - \ln(\ln(1 + 1/\gamma))} \tag{1}$$

where U_{bd} is the breakdown voltage, d is the gap separation, p is the gas pressure, γ is the secondary electron emission coefficient, and A and B are constants determined by the gap type.

While the classical Paschen curve has a right branch with the breakdown voltage decreasing as pd decreases, a characteristic minimum, and a left branch with the breakdown voltage increasing as pd decreases, research has shown that the left branch continues to decrease nearly linearly with d, that is, in microscale gaps, pd scaling fails. Early experiments noted that reducing gap sizes to microscale at atmospheric pressure led to deviations in the traditional breakdown mechanism driven by Townsend avalanche and represented mathematically by Paschen's law (PL) [30]. Departing from the traditional PL, the breakdown voltage would undergo a plateau when the gap width is smaller than ~10 μm and then continue to decrease with the gap width. The gap widths for the transition processes vary with the experimental conditions, such as electrode materials, electrode geometry, applied voltage waveform, gap pressures, etc.

Since the derivation of Paschen's law in the microscale regime was discovered in 1950s, a large number of research work has been dedicated to modification of the classical Paschen's law, from experimental investigation [31, 32] to numerical simulation [33, 34], from atmosphere environment [35] to vacuum [36, 37]. Torres et al. and Slade et al. carried out a series of experimental investigations on microgap breakdown in air and vacuum, respectively. They both found out the plateau stage in the modified Paschen's law and the transition point of gap widths was 4 μm. Besides the numerical simulation, analytic and theoretical calculations have been also carried out. Go [38], Klas [39, 40], Buendia [41], and Loveless [42] calculated the breakdown thresholds at microscale coupling with field emission and Townsend avalanche, considering the ion-enhance field emission, where the electron collision ionization coefficient α and the secondary electron emission coefficient γ dictate the breakdown process, where the secondary electron emission coefficient γ would be enhanced by the space charge accumulation and the cathode charge production through secondary emission. Therefore, the investigation of electrical breakdown behaviors at microscale, including the methodology, fundamental properties, influencing factors, and physical mechanisms, is urgently demanded, which is of critical importance not only for the plasma physics community but also for micro-/nanoelectronic industries.

1.3 The main chapter content

This chapter summarizes the state-of-the-art methodologies, influencing factors, dynamics, and physical mechanisms of the electrical breakdown in microscale based on the research work in the last two decades.

Section 2 summarizes the methodology for investigating the electrical breakdown in microgaps. Section 3 summarizes the influencing factors of the electrical breakdown in microgaps. Section 4 and 5 summarize the dynamic process and physical mechanism of the electrical breakdown in microgaps. Summary and outlook are provided in Section 6.

2. Methodology

Different from the routine gas breakdown experiments in large gaps (>0.1 mm), the electrical breakdown experiments in microgaps (<0.1 mm) require a much better spatial resolution in terms of both observation and gap adjustment. Accordingly, the methodology is very diverse, including the macro electrode structure prepared by the mechanical technique, the planar electrode structure and MEMS device structure prepared by the microfabrication technique, and the microelectrode structure prepared by the electrochemical etching technique. Moreover, the in-situ electro-optical measurement technique has also been proposed for exploring the breakdown dynamic process at microscale. In this chapter, various experimental methods are discussed and summarized.

2.1 The macroelectrode structure

At the initial stage, the study was basically conducted with the macroelectrode structure and experimental setup similar to that at macroscale. **Figure 1a** shows the schematic diagram of a typical macro electrode-based experimental setup and (b) shows the picture of a spherical electrode-based experimental setup used in the literature. The electrode size is in the order of millimeters in radius, which could be fabricated by mechanical machining. One electrode is fixed with the base (also known as static electrode) and the other is movable with the screw micrometer or stepping motor (also known as movable electrode). Both electrodes are required to be aligned on a straight line to ensure the consistence of the discharge experiments.

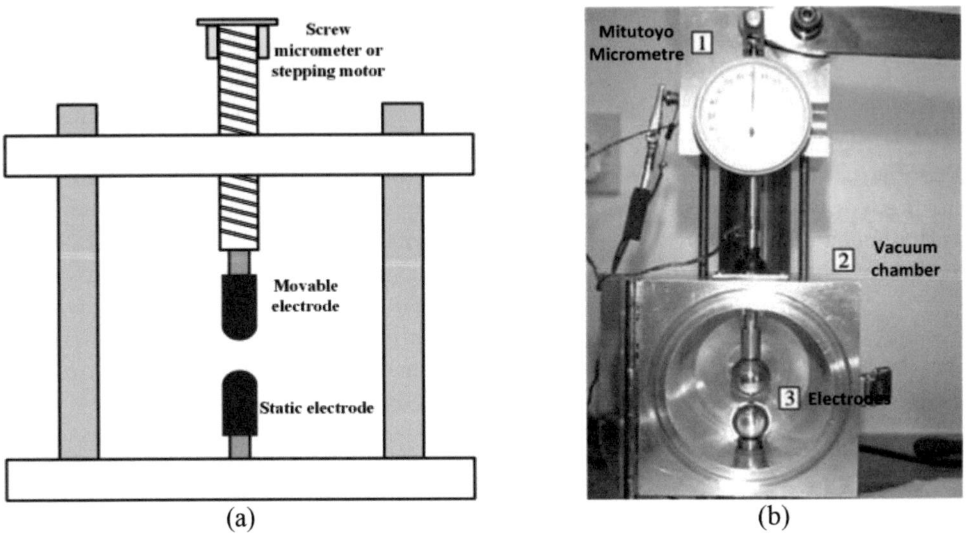

Figure 1.
(a) Schematic diagram of macroelectrode-based experimental setup. The gap distance is controlled and adjusted by the screw micrometer or stepping motor [43]; (b) the picture of two spherical electrode experimental setup for vacuum breakdown test [44].

Therefore, the gap distance could be controlled by adjusting the screw micrometer or stepping motor, with an accuracy of 2 μm. Therefore, this method applies for the electrical breakdown in microgaps ranging from 5 to 500 μm.

2.2 The planar electrode structure

The emerging of microelectronic devices drew intensive attention to the electrical reliability issues, and thus, the planar electrode structure was proposed. Through the standard fabrication process, such as oxidation, lithography, deposition, etching, etc., the planar metal electrode (aluminum, copper, gold, and platinum) is patterned on the silicon dioxide/silicon substrate with a thickness of several hundreds of nanometers and a gap distance ranging from several nanometers to micrometers.

(a) shows the typical planar electrode-based experimental setup. The semicircular type electrode pattern was fabricated on the substrates and the electrical breakdown experiments could be conducted between microgaps. In addition, the suspended planar electrode was also proposed by sacrificing layer process as shown in **Figure 2b**, in which electrical breakdown properties of MEMS devices (such as MEMS switches and MEMS motors) could be investigated. Therefore, this method is dedicated to the study of device reliability issues with typical and simple structures.

2.3 The MEMS device structure

Apart from the typical simplified electrode structures above, lots of research work has also focused on the breakdown characterization and reliability assessment of real device structures under ESD impact, especially for those devices that require electrostatic actuation (i.e., RF MEMS switch, micro-motor, and micro-mirror) or are very susceptible to static charge accumulation (i.e., photomask). **Figure 3** shows the pictures of different RF MEMS devices for ESD impact testing, which are gold-based capacitive (a) and ohmic (b–d) RF-MEMS switches with vertical air-gap structure from 1.0 to 4.5 μm and lateral air-gap structure of 6.7 μm. For this

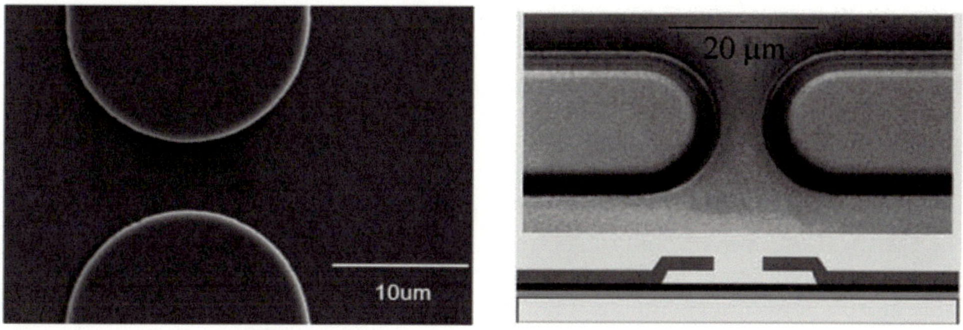

(a) (b)

Figure 2.
Schematic diagram of planar electrode-based experimental setup: (a) semicircular type electrode [45] and (b) suspended semicircular type electrode [46].

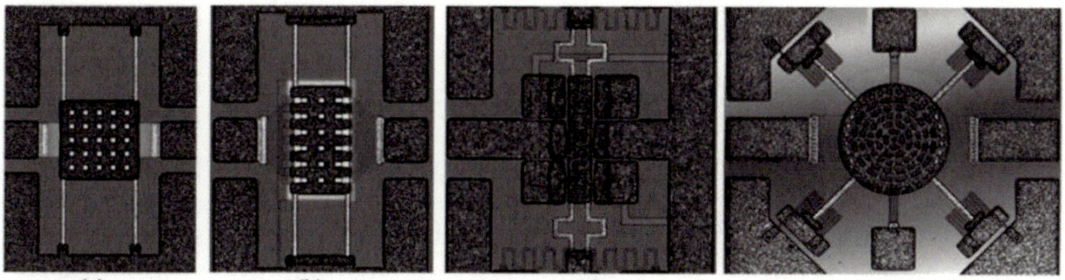

(a) (b) (c) (d)

Figure 3.
Tested devices were gold-based (a) capacitive and (b–d) ohmic RF-MEMS switches [10].

configuration, the breakdown may occur across the micron air gaps of RF MEMS switches and result in permanent physical damage on the devices.

Figure 4a shows the SEM image of a torsional ratcheting actuator (TRA) in which the ratchet gear and curved comb fingers are used for electrostatic actuation and (b) shows the optical image of metal-air-metal device on reticle with 4 μm gap,

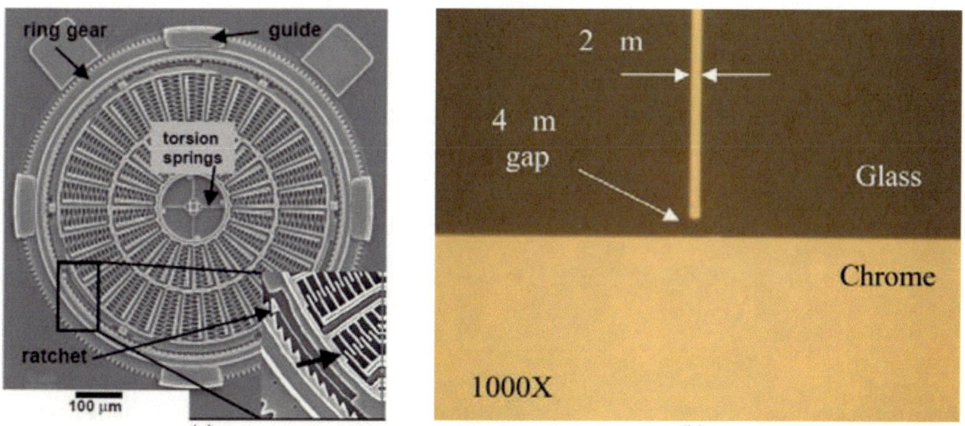

(a) (b)

Figure 4.
(a) SEM image of a torsional ratcheting actuator (TRA). The inset shows an enlarged view of the ratchet gear and curved comb fingers used for electrostatic actuation [15]; (b) optical image of metal-air-metal device on reticle with 4 μm gap [7].

which has been developed to check the ESD threat to reticles in a photolithography bay. For this configuration, the breakdown may occur across the surface of the air-gap structure and result in permanent physical damage on the devices.

2.4 The microelectrode structure

While the planar electrode and MEMS device structure are employed to explore the electrical reliability of microelectronic devices, the intrinsic properties of electrical breakdown in microgaps require microelectrodes with precisely controllable morphology and geometry, which were proposed and fabricated by combining the electrochemical etching and Joule melting method [47]. **Figure 5** shows the microelectrode structure-based experimental setup, of which the hemisphere electrodes were made of tungsten, and the radius of the electrodes ranged from 50 nm to 200 μm. The hemisphere electrodes have a regular and contaminant-free surface. The three-dimensional piezoelectric displacement could align the electrode pair with the aid of an optical microscope, allowing precise gap adjustment from 1 to 25 μm with an uncertainty of ±100 nm.

2.5 The in-situ electro-optical experimental setup

Basically, the fundamental properties and influencing factors of the electrical breakdown in microgaps could be obtained by measuring the electrical parameters; however, to further understand the dynamics and physical mechanism, additional physical parameters during the breakdown are required. Monitoring the optical properties of the breakdown dynamic process is the primary way, which may need to satisfy two requirements simultaneously: (1) how to observe the breakdown channel at microscale and (2) how to capture the breakdown appearance in nanoseconds.

Figure 6 shows the electro-optical measurement setup that can simultaneously fulfill these requirements. The system consists of a nanosecond pulse generation unit, a synchronous and delay triggering unit, an in-situ optical imaging unit, and an electrical parameter measurement unit. The nanosecond pulse generation unit can provide amplitude-adjustable pulses up to 5 kV. The synchronous and delay triggering unit is achieved by a dual-channel function signal generator which can adjust the relative time delay between the two TTL triggering signals and ensure the synchronism of the test. The in-situ optical imaging unit integrates the optical microscope for micron-scale spatial resolution (1 μm) and the high-speed gated ICCD camera for nanosecond-scale temporal resolution (2 ns). The breakdown current and voltage are measured by a current coil (1 A/V) and a voltage attenuator (100:1), and then recorded by a digital oscilloscope. This system allows temporal

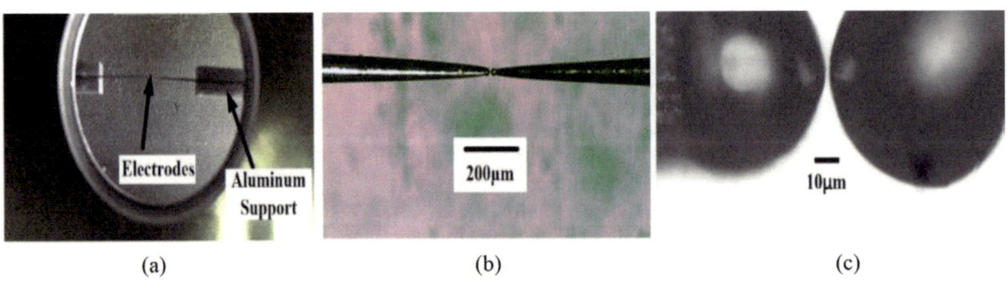

Figure 5.
Schematic diagram of sphere-to-sphere microelectrode-based experimental setup: (a) 10×, (b) 50×, and (c) 1000× magnification [48].

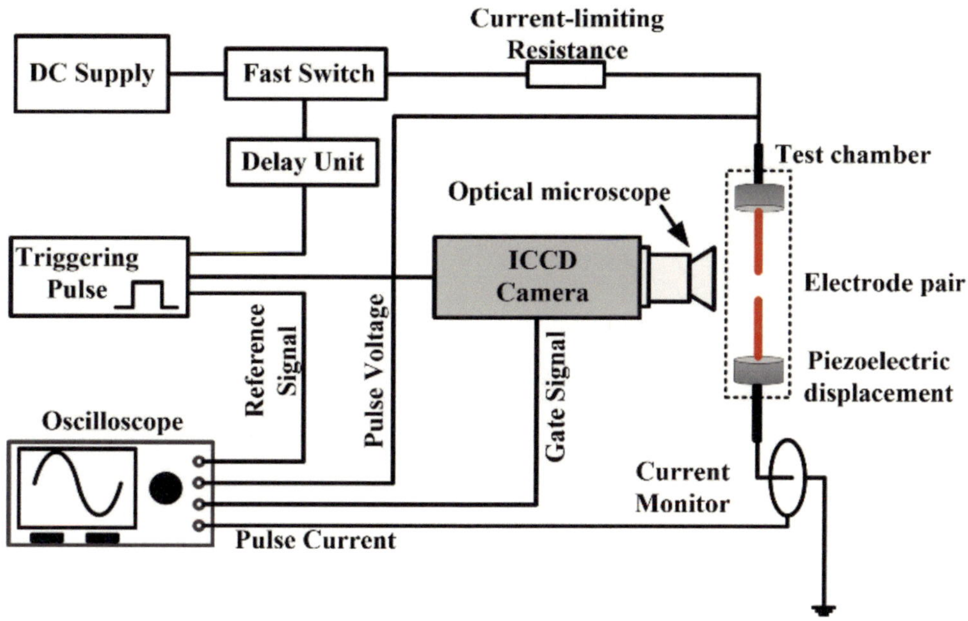

Figure 6.
Schematic diagram of in-situ electro-optical measurement system [48, 49].

and spatial-resolved optical measurement and images the discharge appearance of pulse breakdown across microgaps, which will be a promising method to further explore the underlying principle of gas breakdown at microscale and evaluate the insulation performance in micro-/nanoelectronics.

3. Influencing factors of electrical breakdown in microgaps

As the gap size decreases, the classical Paschen's law demonstrates a significant derivation at microscale which implies the different physical mechanisms from Townsend avalanche breakdown. Since a lot of influencing factors could affect the electrical breakdown, this section gives some of the influencing factors such as the gap widths, the atmospheric pressures, and the applied voltages. These results determine quantitative relationships between the breakdown and the factors, and thus provide an overall picture of the electrical breakdown in microgaps.

3.1 The effect of the gap widths

Figure 7 shows the breakdown thresholds as a function of gap width in atmospheric air (101 kPa) at room temperature (298.15 K). The electrode configuration is hemisphere-hemisphere with gap widths from 1 to 25 μm. For gap widths <5 μm, the breakdown voltage decreases with decreasing gap width. For gap widths between 5 and 10 μm, the breakdown voltages almost remain constant at about 490V regardless of the gap width, demonstrating a "plateau" stage. Although numerous microscale breakdown studies have noted this plateau [50], a strong hypothesis has not yet been developed. For gap widths larger than 10 μm, breakdown voltage increases dramatically with increasing gap width, indicating the increasing importance of Townsend avalanche. It can be noted that the breakdown voltage is 386 V when the gap width is 1 μm and the breakdown voltage is 842 V when the gap width is 25 μm. As the gap width shrinks to several micrometers, the

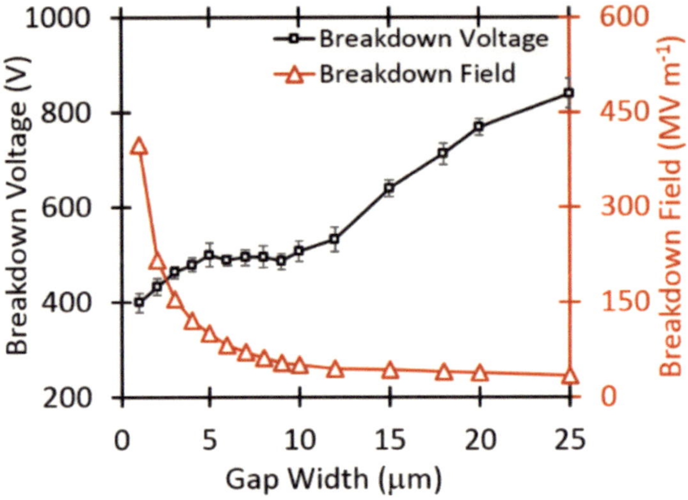

Figure 7.
Measured breakdown voltage and electric field as a function of gap widths, the error bars represent the standard deviation of the measured breakdown voltage.

number of gas molecules inside the gap would be not enough for impact ionization, thus higher field strength is demanded for electron avalanche. When the gap width is reduced to <5 μm, the electric field strength is calculated to be ~10^8 V/m, which has reached the threshold of field electron emission from the electrode surface. The obvious transition in the curves can be noticed and the cathode field emission plays a dominant role in the generation of free electrons.

3.2 The effect of applied voltages

Figure 8 shows the breakdown thresholds as a function of applied voltages in atmospheric air (760 Torr) and room temperature (298.15 K), the electrode configuration is hemisphere-hemisphere type with various gap widths from 1 to 25 μm, and

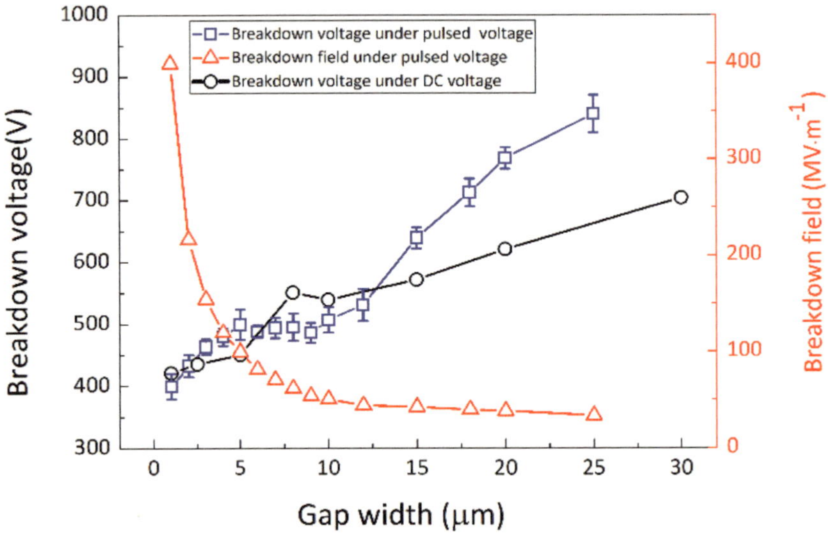

Figure 8.
Breakdown thresholds as a function of gap widths under nanosecond pulsed voltage and DC voltage.

the error bars in all these figures show the standard deviation above and below the mean value of measurement. The results of pulsed breakdown [51] and DC breakdown [32] are plotted for comparisons. Generally, the nanosecond pulsed breakdown thresholds are at least two or three times higher than DC breakdown [52]; however, an interesting phenomenon can be observed from **Figure 8** when the gap width is scaled down to 15 μm. It can be seen that, overall, Actually, a lot of numerical under nanosecond pulsed voltage (blue solid square) shows a similar trend and amplitude to those under DC voltage (black solid square). For a 15-μm gap, the pulsed breakdown voltage is 639 V while the DC breakdown voltage is 571 V. For a 5-μm gap, the pulsed breakdown voltage is 450 V while the DC breakdown voltage is 499 V. More specifically, it can be noted that there is also a "plateau" stage between 5 and 10 μm, with a constant breakdown voltage of about 490 V, which is considered to be the transition region from Townsend avalanche to ion-enhanced field emission.

When the gap width is <5 μm, the breakdown voltage decreases with the decrease of gap width, demonstrating a good consistence with the DC breakdown voltage ($U_{\text{pulsed}} = 432$ V $\approx U_{\text{DC}} = 435$ V for the 3-μm gap). Meanwhile, the pulsed breakdown voltage is found to have a power law dependence on the gap width through conducting the fitting analysis: $U = 396 \times b^{0.14}$, where U is the breakdown voltage in Volt, b is the gap width in micrometer and the Adj. R-Square is 0.99195. That is in good agreement with the vacuum breakdown behaviors proposed by Staprans in 1966 [53, 54], implying that while the gap width is reduced to 5 μm, the pulsed breakdown in air might be similar to the vacuum breakdown. So as the gap width shrinks to several micrometers, the number of gas molecules inside the gap would be not enough for the collision ionization, and thus, higher field strength is demanded for electron avalanche. When the gap width is reduced to <5 μm, the electric field strength is calculated to be 108 V/m, which has reached the threshold of field electron emission from the electrode surface. The obvious transition in the curves can be noticed and the cathode field emission is believed to play a dominant role in the generation of free electrons.

However, as the gap width continues to increase from 15 μm, it is noteworthy that pulsed breakdown demonstrates larger thresholds, and furthermore exhibits a linear increase with a positive slope of 21.5 compared to the positive slope of 8.4 for the DC breakdown voltages, which indicates that the duration of applied voltage determines the amplitude of breakdown voltage [55], that is, the breakdown under the nanosecond pulse would be much more difficult to breakdown than the DC voltage, and thus, it could be expected to become two or three times larger than the DC breakdown values as mentioned above.

3.3 The effect of atmospheric pressures

Figure 9 shows the breakdown thresholds as a function of atmospheric pressures [56], the squares represent the breakdown thresholds at a pressure of 760 Torr, the circles represent the breakdown thresholds at a pressure of 375 Torr, and the triangles represent the breakdown thresholds at a pressure of 23 Torr. The electrode configuration is hemisphere-hemisphere type with various gap widths from 1 to 25 μm. Apparently, the curves demonstrate a similar trend; however, the breakdown voltages are almost the same when the gap width is <5 μm. Considering an electron as the gas molecule, the mean free path of an electron $\acute{\lambda}_e$ can be derived from the following equation.

$$\acute{\lambda}_e = \frac{K_B \cdot T}{\pi\left(\frac{d_m + d_e}{2}\right)^2 \cdot p} \tag{2}$$

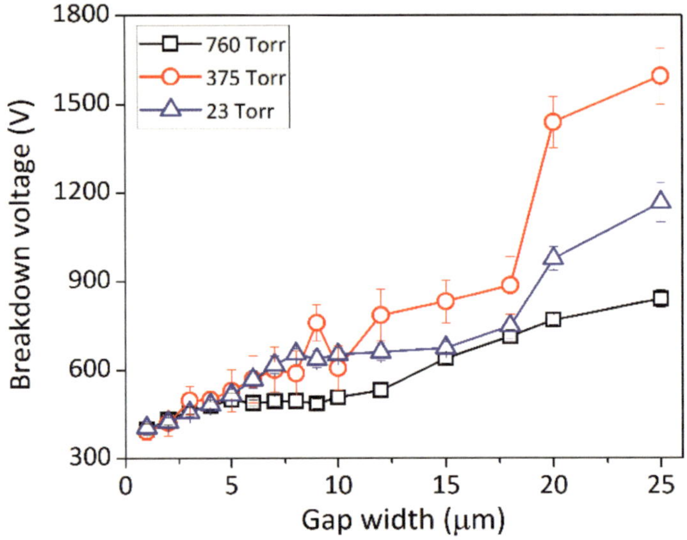

Figure 9.
Breakdown voltages as a function of gas pressure at different pressures.

where K_B is the Boltzmann constant (1.38×10^{-23} J/K), T is the ambient temperature in Kelvins, d_m is the atom or molecular diameter in meter, d_e is the electron diameter in meter, and p is the atmospheric pressure in Pascal. Since d_e is 5.62×10^{-15} m which is <1/1000 of the proton diameter, the collision between electrons could be neglected, so the mean free path of an electron could be defined to be the average distance the electron travels between successive collisions with the gas atom or molecule, that is, $\lambda_e = \frac{K_B \cdot T}{\pi\left(\frac{d_m}{2}\right)^2 \cdot p}$, which is inversely proportional to the square of the gas molecule diameter [57]. According to Eq. (2), when the gas pressure is 760, 375, and 23 Torr, the mean free path of an electron in air is calculated to be 539 nm, 1.1 μm, and 18 μm, respectively, which are either much smaller or comparable with the gap length (5 μm), so the moving electrons can seldom collide with the gas molecules in the gap space and the number of collisions is so small that no considerable electrons and ions could be produced, in other words, the gaseous gap is almost equivalent to vacuum gap at this scale. As the gap width increases, sufficient and more collision ionization can take place at 760 Torr than those at 23 and 375 Torr due to larger propagation distances (>5 μm), which result in the significant difference between the breakdown thresholds. This implies that the role of gas molecule density or atmospheric pressure inside the gap could be eliminated when the gap width is <5 μm but will greatly affect the breakdown process in larger gaps. However, it also demonstrates a different trend that the breakdown thresholds at 375 Torr are larger than those at 23 Torr, which will be further investigated in the future study.

4. The dynamics of electrical breakdown in microgaps

Except for the fundamental properties of electrical breakdown in microgaps, the breakdown evolution process was also investigated for further understanding the dynamics properties, with the aid of the in-situ electro-optical measurement system introduced in Section 2.5. This section provides the temporal evolution of gas breakdown which exhibits various breakdown channel morphologies and transitions dependent upon the gap width, and highlights the breakdown dynamics in microgaps.

4.1 The breakdown paths

Figure 10 shows the breakdown morphology and discharge paths for various gap widths ranging from 1 to 20 μm. The sphere-sphere electrodes are employed in the atmospheric air environment, and the triggering time of the ICCD shutter is 10 μs prior to the breakdown moment with an exposure time of 200 ms, which guarantees that the entire breakdown process could be captured and recorded within one shot. It can be seen from **Figure 10a–c** with a gap width of 20, 15, and 12 μm, that the luminescence fills the entire gap and surroundings, in which an intense light channel can be clearly observed between the electrodes. Typically, the discharge plasma would propagate along the shortest distance between the electrodes, and the spot with maximum electric field strength is at the apex of the sphere electrodes, so the straight line connecting the apexes is considered to be the shortest path for the breakdown, which could be proved by the captured images. However, an interesting phenomenon is observed in **Figure 10d–f** with a gap width of 9, 7, and 5 μm, that the intense light channel does not follow the very straight line between the electrodes; on the contrary, it initiates from the cathode apex and propagates along a curved line

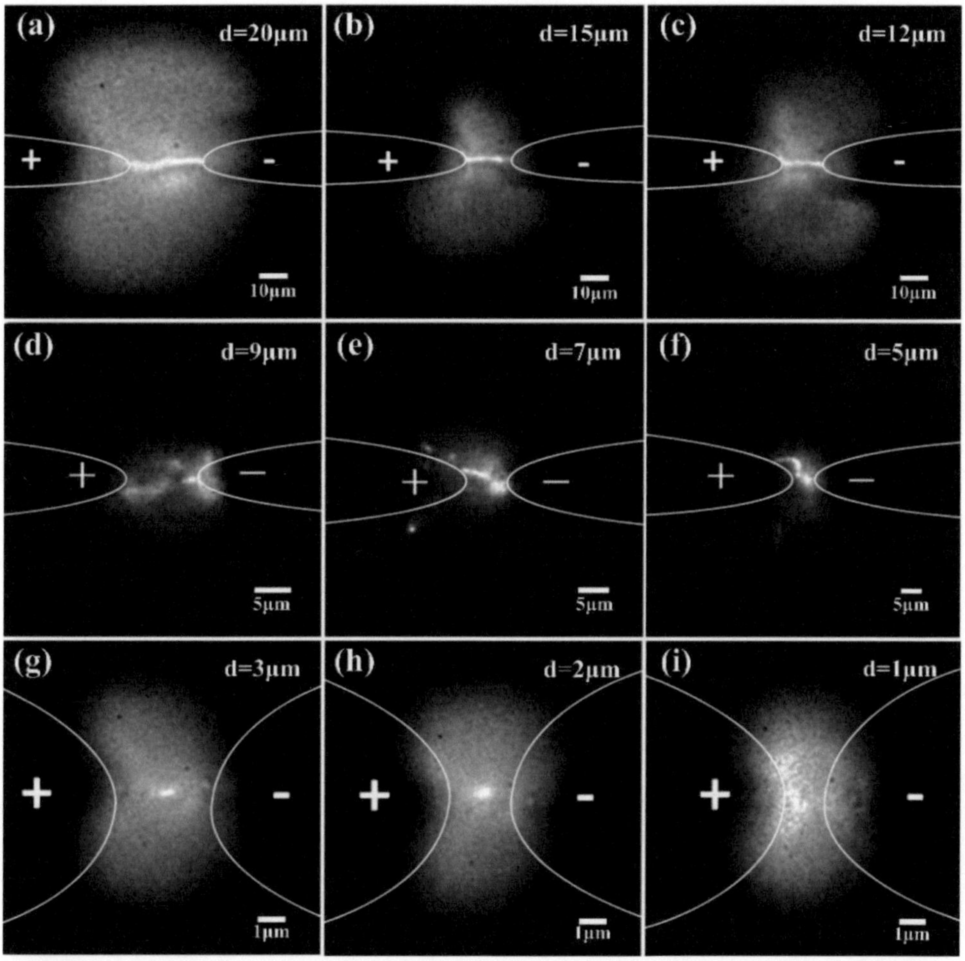

Figure 10.
Breakdown morphology at gap widths from 1 to 20 μm. (a–c) show the breakdown propagating along the shortest path with luminescence filling the surrounding area, (d–f) show the roughly constant path lengths regardless of gap width which is consistent with the plateau of breakdown voltage in this region, and (g–i) indicate no obvious breakdown channel arising at these smallest gap distances [48].

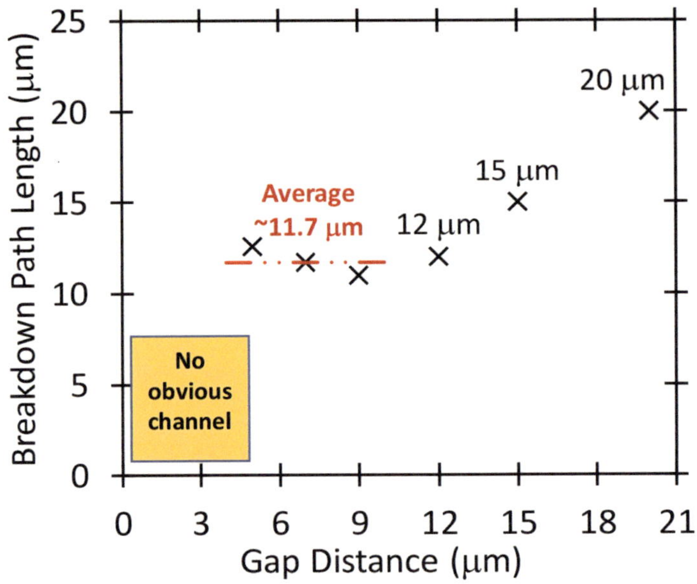

Figure 11.
The effective lengths of breakdown path for various gap widths [48].

to the neighbor region of the anode apex, which is a significant deviation from the theoretical prediction. In **Figure 10g–i** with a gap width of 3, 2, and 1 μm, the entire gap is full of luminescence and no obvious breakdown channel could be observed. While a channel may arise for the 2 and 3 μm gaps, it is much fainter compared to overall luminous intensity of the remainder of the diffuse discharge, unlike the noticeably higher intensity channels that connect both electrodes at larger gaps [48].

Figure 11 shows the effective lengths of breakdown paths in different gaps according to the breakdown channel images in **Figure 10**. It can be noteworthy that the curved path in **Figure 10d–f** is almost the same (about 11.7 μm) regardless of the gap widths, which is well consistent with the trend of the breakdown voltages in **Figure 7** and would be a very straight evidence to explain the "plateau" stage from 5 to 10 μm. That is, the consistency between the plateau in breakdown voltage and the constant breakdown path length for gap widths ranging from 5 to 10 μm is critical for understanding the transition in breakdown mechanism both experimentally and theoretically. It implies that the extension of breakdown path provides more collision ionization and electron avalanches for the breakdown which means that the ion-enhanced field emission must play an important role in breakdown rather than the Townsend avalanche alone, thus resulting in the "plateau" stage. Therefore, this evidence directly shows the transition from Townsend avalanche to ion-enhanced field emission, in which the field emission begins to dominate over Townsend avalanche for gaps smaller than ~10 μm and Townsend avalanche becomes continuously less important for smaller gaps, and finally, the field emission will dominate the breakdown for gaps shorter than 5 μm [48].

5. The physical mechanism of electrical breakdown in microgaps

Based on the captured breakdown morphology across various microgaps, the physical mechanisms could be summarized as follows:

a. When the gap width d is larger than 10 μm, the breakdown threshold is expressed as a function of the product gas pressure p and gap width d, and the

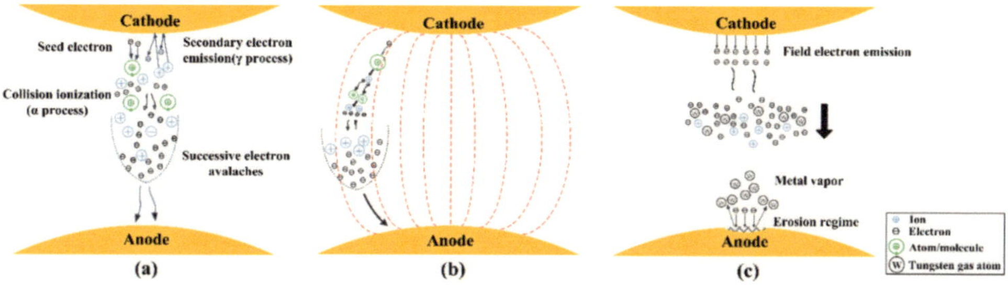

Figure 12.
The physical process unifying Townsend avalanche and field emission for microscale breakdown for (a) d > 10 μm, (b) d = 5–10 μm, and (c) d = 1–5 μm [48].

Townsend avalanche is considered to be the dominant mechanism. The breakdown demonstrates a clear electron avalanche plasma trajectory connecting the cathode tip and the anode tip by a straight path, as shown in **Figure 12a**.

b. When the gap width *d* lies between 10 and 5 μm, Townsend avalanche still plays a role in breakdown but the contribution of ion enhanced field emission becomes more important. In this regime, a plateau can be observed indicating that the breakdown thresholds almost remain invariant as the decease of the gap width, which shows the transient from Townsend avalanche to the field emission process. Although the gap length is not long enough for the collision ionization, the initial electron avalanche is generated in the vicinity of the cathode and propagates along a curved path following the electric field lines. This could extend the effective propagation width and then may increase the collision ionization probability and frequency. The successive electron avalanches would be produced and may ultimately contribute to inducing breakdown, as shown in **Figure 12b**.

c. When the gap width *d* is smaller than 5 μm, the breakdown threshold demonstrates a linear relationship with the gap width. In this regime, a high electric field (~10^8 V/m) would reduce the potential barrier of the cathode and electrons would be emitted into the gap, so the initial electron avalanche can be generated around the cathode tip. Since the electron mean free path is comparable to the gap length, the emitted electrons would drift toward the anode and collide with the anode directly, resulting in the heating and release of anode and cathode materials due to the Nottingham effect. Then the thermal electron emission would turn on and more electrons would be generated by the combination of field emission and thermal emission. The outgas and atoms would fill the gap, and finally breakdown would occur with a steep decline of predicted voltage thresholds, which indicates that field emission is the dominated mechanism for gap width <5 μm, as shown in **Figure 12c**.

6. Summary and outlook

This chapter provides a general review of the electrical breakdown in microgaps, including the methodology, influencing factors, dynamics, and physical mechanism. The breakdown thresholds in various conditions and the transition from Townsend avalanche to field emission-driven breakdown were demonstrated, which would be vital to the electrical breakdown theory at microscale. Meanwhile, understanding the fundamental mechanism of gas breakdown at microscale will have far reaching impact on practical devices due to the numerous applications that leverage microplasmas [21], including excimer lamps with emissions in the VUV [58], ozone generators [59], arrays for flat panel light sources [60], nanoparticle synthesis [61], medicine [62], environmental remediation [63], detectors [64, 65], microthrusters [66], and combustion [67]. While a lot of numerical calculation work devoted to this subject could be found in somewhere else, this chapter focuses on the experimental investigations of breakdown behaviors in microgaps, which helps to pave the way for insulation design and discharge applications at small scales.

As the miniaturization trend of devices and equipment continues along with the great demand in civil and military industries, the electrostatic sensitivity increases accordingly, leading to a new failure mechanism [26]. When the physical size downscales to nanoscale and molecular scale, the quantum effect, space charge effect, and other effects should be considered, and this will also require novel experimental techniques that can obtain more physical parameters during the breakdown process. Therefore, with advanced experimental techniques, more and more explorations in breakdown behaviors at microscale, nanoscale, and molecular scale will surely be carried out, and new physical mechanisms will be put forward in the future.

Acknowledgements

This chapter was supported by the National Natural Science Foundation of China (Grant No. 51607138, 51521065), the Fundamental Research Funds for the Central Universities (Grant No. xzy012019030), and China Postdoctoral Science Foundation (Grant No. 2016M602820).

Conflict of interest

The authors declare that there is no conflict of interest regarding the publication of this chapter.

Author details

Guodong Meng* and Yonghong Cheng
Xi'an Jiaotong University, Xi'an, China

*Address all correspondence to: gdmengxjtu@xjtu.edu.cn

References

[1] Brock DC, Moore GE. Understanding Moore's Law: Four Decades of Innovation. San Francisco: Chemical Heritage Foundation; 2006. DOI: 10.5860/choice.44-6278

[2] Rebeiz GM, Muldavin JB. RF MEMS switches and switch circuits. IEEE Microwave Magazine. 2001;**2**(4):59-71. DOI: 10.1109/6668.969936

[3] Bilici MA, Haase JR, Boyle CR, Go DB, Sankaran RM. The smooth transition from field emission to a self-sustained plasma in microscale electrode gaps at atmospheric pressure. Journal of Applied Physics. 2016;**119**(22):223301. DOI: 10.1063/1.4953648

[4] Steinman A. Reducing electrostatic related defects in photolithography. In: IEEE International Symposium on Semiconductor Manufacturing; 1994. DOI: 10.1109/ISSM.1994.729449

[5] Montoya J, Levit L, Englisch A. A study of the mechanisms for ESD damage to reticles. IEEE Transactions on Electronics Packaging Manufacturing. 2001;**24**(2):78-85. DOI: 10.1109/6104.930957

[6] Wallash A. A study of shunt ESD protection for GMR recording heads. Journal of Electrostatics. 2002;**56**(3):295-302. DOI: 10.1016/S0304-3886(02)00091-8

[7] Wallash A, Levit L. Electrical breakdown and ESD phenomena for devices with nanometer-to-micron gaps. In: Reliability, Testing, and Characterization of MEMS/MOEMS II; 2003. pp. 87-96. DOI: 10.1117/12.478191

[8] Wallash A et al. Electrostatic discharge testing of tunneling magnetoresistive (TMR) devices. IEEE Transactions on Magnetics. 2000;**36**(5):2809-2811. DOI: 10.1109/20.908596

[9] Voldman SH. ESD Basics: From Semiconductor Manufacturing to Product Use. Chichester: John Wiley & Sons; 2012. DOI:10.1002/9781118443323

[10] Tazzoli A et al. A comprehensive study of MEMS behavior under EOS/ESD events: Breakdown characterization, dielectric charging, and realistic cures. In: IEEE Electrical Overstress/Electrostatic Discharge Symposium Proceedings; 2010. DOI: 10.1016/j.elstat.2011.07.007

[11] Sangameswaran S et al. A study of breakdown mechanisms in electrostatic actuators using mechanical response under EOS-ESD stress. In: IEEE 31st EOS/ESD Symposium; 2009

[12] Tazzoli A, et al. EOS/ESD sensitivity of functional RF-MEMS switches. In: IEEE 30th Electrical Overstress/Electrostatic Discharge Symposium; 2008

[13] Tazzoli A, Peretti V, Meneghesso G. Electrostatic discharge and cycling effects on ohmic and capacitive RF-MEMS switches. IEEE Transactions on Device and Materials Reliability. 2007;**7**(3):429-437. DOI: 10.1109/tdmr.2007.907422

[14] Walraven JA, et al. Electrostatic discharge/electrical overstress susceptibility in MEMS: A new failure mode. In: Proceedings SPIE 4180, MEMS Reliability for Critical Applications; 2000. pp. 30-39. DOI: 10.1117/12.395703

[15] Walraven JA, et al. Human body model, machine model, and charge device model ESD testing of surface micro machined microelectromechanical systems (MEMS). In: IEEE Electrical Overstress/Electrostatic Discharge Symposium; 2001

[16] Sangameswaran S, et al. ESD reliability issues in

microelectromechanical systems (MEMS): A case study on micromirrors. In: IEEE 30th Electrical Overstress/ Electrostatic Discharge Symposium; 2008

[17] Sangameswaran S et al. Impact of design factors and environment on the ESD sensitivity of MEMS micromirrors. Microelectronics Reliability. 2010;**50** (9-11):1383-1387. DOI: 10.1016/j. microrel.2010.07.079

[18] Cheng Y, Meng G, You XG, Chen L, Wu K, Pan C. Experimental study on electrical breakdown of micro-electro-mechanical systems with micro-gaps. In: 2012 Annual Report Conference on Electrical Insulation and Dielectric Phenomena; 2012. pp. 339-342. DOI: 10.1109/CEIDP.2012.6378790

[19] Steinman A, Henry LG, Hernandez M. Measurements to establish process ESD compatibility. In: IEEE Electrical Overstress/Electrostatic Discharge Symposium; 2010

[20] Schoenbach KH, Becker K. 20 years of microplasma research: A status report. The European Physical Journal D. 2016;**70**(2):29. DOI: 10.1140/epjd/ e2015-60618-1

[21] Becker K. Microplasmas, a platform technology for a plethora of plasma applications. The European Physical Journal Special Topics. 2017;**226**(13):2853-2858. DOI: 10.1140/ epjst/e2016-60375-4

[22] Charles C, Bish A, Boswell R, Dedrick J, Greig A, Hawkins R, et al. A short review of experimental and computational diagnostics for radiofrequency plasma micro-thrusters. Plasma Chemistry and Plasma Processing. 2016;**36**(1):29-44. DOI: 10.1007/s11090-015-9654-5

[23] Wright W, Ferrer P. Electric micropropulsion systems. Progress in Aerospace Sciences. 2015;**74**:48-61. DOI: 10.1016/j.paerosci.2014.10.003

[24] Levchenko I, Xu S, Teel G, Mariotti D, Walker M, Keidar M. Recent progress and perspectives of space electric propulsion systems based on smart nanomaterials. Nature Communications. 2018;**9**(1):879. DOI: 10.1038/s41467-017-02269-7

[25] Martinez-Sanchez M, Pollard JE. Spacecraft electric propulsion-an overview. Journal of Propulsion and Power. 1998;**14**(5):688-699. DOI: 10.2514/2.5331

[26] Voldman SH. Nano electrostatic discharge. IEEE Nanotechnology Magazine. 2009;**3**(3):12-15. DOI: 10.1109/MNANO.2009.934212

[27] Voldman SH. Electrical Overstress (EOS): Devices, Circuits and Systems. Chichester: John Wiley & Sons; 2013. DOI: 10.1002/9781118511886

[28] Voldman SH. ESD: Failure Mechanisms and Models. Chichester: John Wiley & Sons; 2009. DOI:10.1002/9780470747254

[29] Paschen F. On sparking over in air, hydrogen, carbon dioxide under the potentials corresponding to various pressures. Wiedemann Annalen der Physik und Chemie. 1889;**37**:69-96

[30] Boyle W, Kisliuk P. Departure from Paschen's law of breakdown in gases. Physical Review. 1955;**97**(2):255-259. DOI: 10.1103/PhysRev.97.255

[31] Iwabuchi H, Morimoto T, Matsuoka S, Kumada A, Hidaka K. Pre-breakdown phenomenon in micrometer-scale gap. In: 31st ICPIG; 2013

[32] Radmilović-Radjenović M, Matejčik Š, Klas M, Radjenović B. The role of the field emission effect in direct-current argon discharges for the gaps ranging from 1 to 100 μm. Journal of Physics D: Applied Physics. 2012;**46**(1):015302. DOI: 10.1088/0022-3727/46/1/015302

[33] Li Y, Tirumala R, Rumbach P, Go DB. The coupling of ion-enhanced field emission and the discharge during microscale breakdown at moderately high pressures. IEEE Transactions on Plasma Science. 2013;**41**(1):24-35. DOI: 10.1109/TPS.2012.2224380

[34] Venkattraman A, Garg A, Peroulis D, Alexeenko AA. Direct measurements and numerical simulations of gas charging in microelectromechanical system capacitive switches. Applied Physics Letters. 2012;**100**(8):083503. DOI: 10.1063/1.3688176

[35] Venkattraman A, Alexeenko AA. Scaling law for direct current field emission-driven microscale gas breakdown. Physics of Plasmas. 2012;**19**(12):123515. DOI: 10.1063/1.4773399

[36] Marić R, Stanković K, Vujisić M, Osmokrović P. Electrical breakdown mechanisms in vacuum diodes. Vacuum. 2010;**84**(11):1291-1295. DOI: 10.12693/APhysPolA.118.585

[37] Sudarshan T, Ma X, Muzykov P. High field insulation relevant to vacuum microelectronic devices. IEEE Transactions on Dielectrics and Electrical Insulation. 2002;**9**(2):216-225. DOI: 10.1109/94.993738

[38] Go DB, Pohlman DA. A mathematical model of the modified Paschen's curve for breakdown in microscale gaps. Journal of Applied Physics. 2010;**107**(10):103303. DOI: 10.1063/1.3380855

[39] Klas M, Matejcik S, Radjenovic B, Radmilovic-Radjenovic M. Experimental and theoretical studies of the direct-current breakdown voltage in argon at micrometer separations. Physica Scripta. 2011;**83**(4):045503. DOI: 10.1088/0031-8949/83/04/045503

[40] Klas M, Matejcik S, Radjenovic B, Radmilovic-Radjenovic M. Experimental and theoretical studies of the breakdown voltage characteristics at micrometre separations in air. Europhysics Letters. 2011;**95**(3):35002. DOI: 10.1209/0295-5075/95/35002

[41] Buendia JA, Venkattraman A. Field enhancement factor dependence on electric field and implications on microscale gas breakdown: Theory and experimental interpretation. EPL (Europhysics Letters). 2015;**112**(5):55002. DOI: 10.1209/0295-5075/112/55002

[42] Loveless AM, Garner AL. A universal theory for gas breakdown from microscale to the classical Paschen law. Physics of Plasmas. 2017;**24**(11):113522. DOI: 10.1063/1.5004654

[43] Cheng Y, Meng G, Dong C. Review on the breakdown characteristics and discharge behaviors at the micro&nano scale. Transactions of China Electrotechnical Society. 2017;**32**(2):13-23. DOI: CNKI:SUN:DGJS.0.2017-02-002

[44] Torres JM, Dhariwal RS. Electric field breakdown at micrometer separations in air and vacuum. Microsystem Technologies. 1999;**6**(1): 6-10. DOI: 10.1007/s005420050166

[45] Meng G, Cheng Y, Gao X, Wang K, Dong C, Zhu B. In-situ optical observation of dynamic breakdown process across microgaps at atmospheric pressure. IEEE Transactions on Dielectrics and Electrical Insulation. 2018;**25**(4):1502-1507. DOI: 10.1109/TDEI.2018.007017

[46] Strong F, Skinner JL, Tien NC. Electrical discharge across micrometerscale gaps for planar MEMS structures in air at atmospheric pressure. Journal of Micromechanics and Microengineering. 2008;**18**(7):075025. DOI: 10.1088/0960-1317/18/7/075025

[47] Meng G, Cheng Y, Wu K, Chen L. Electrical characteristics of nanometer gaps in vacuum under direct voltage. IEEE Transactions on Dielectrics and Electrical Insulation. 2014;**21**(4): 1950-1956. DOI: 10.1109/TDEI.2014.004376

[48] Meng G, Gao X, Loveless AM, Dong C, Zhang D, Wang K, et al. Demonstration of field emission driven microscale gas breakdown for pulsed voltages using in-situ optical imaging. Physics of Plasmas. 2018;**25**(8):082116. DOI: 10.1063/1.5046335

[49] Meng G, Ying Q, Loveless AM, Wu F, Wang K, Fu Y, et al. Spatio-temporal dynamics of pulsed gas breakdown in microgaps. Physics of Plasmas. 2019;**26**(1):014506. DOI: 10.1063/1.5081009

[50] Go D, Venkattraman A. Microscale gas breakdown: Ion-enhanced field emission and the modified Paschen's curve. Journal of Physics D: Applied Physics. 2014;**47**(50):503001. DOI: 10.1088/0022-3727/47/50/503001

[51] Meng G. Experimental and numerical investigation of the influencing factors on pulsed gas breakdown in microgaps. In: Preparation; 2019

[52] Schnyder R, Howling A, Bommottet D, Hollenstein C. Direct current breakdown in gases for complex geometries from high vacuum to atmospheric pressure. Journal of Physics D: Applied Physics. 2013;**46**(28):285205. DOI: 10.1088/0022-3727/46/28/285205

[53] Lyon D, Hubler A. Gap size dependence of the dielectric strength in nano vacuum gaps. IEEE Transactions on Dielectrics and Electrical Insulation. 2013;**20**:1467-1471. DOI: 10.1109/TDEI.2013.6571470

[54] Staprans A. Voltage breakdown limitations of electron guns for high power microwave tubes. In: Proceedings of the Second International Symposium Insulation of High Voltages Vacuum; 1966. pp. 293-303

[55] Tao S, Guangsheng S, Ping Y, Jue W, Weiqun Y, Yaohong S, et al. An experimental investigation of repetitive nanosecond-pulse breakdown in air. Journal of Physics D: Applied Physics. 2006;**39**(10):2192. DOI: 10.1088/0022-3727/39/10/030

[56] Loveless AM, Meng G, Ying Q, Wu F, Wang K, Cheng Y, et al. The transition to Paschen's law for microscale gas breakdown at subatmospheric pressure. Scientific Reports. 2019;**9**(1):5669. DOI: 10.1038/s41598-019-42111-2

[57] Compton KT. The mean free path of an electron in a gas and its minimum ionizing potential. Physical Review. 1916;**8**(4):386. DOI: 10.1103/PhysRev.8.386

[58] Kogelschatz U. Ultraviolet excimer radiation from nonequilibrium gas discharges and its application in photophysics, photochemistry and photobiology. Journal of Optical Technology. 2012;**79**(8):484-493. DOI: 10.1364/JOT.79.000484

[59] Kim K, Park S, Eden J. Self-patterned aluminium interconnects and ring electrodes for arrays of microcavity plasma devices encapsulated in Al_2O_3. Journal of Physics D: Applied Physics. 2007;**41**(1):012004. DOI: 10.1088/0022-3727/41/1/012004

[60] Eun CK, Gianchandani YB. Microdischarge-based sensors and actuators for portable microsystems: Selected examples. IEEE Journal of Quantum Electronics. 2012;**48**(6): 814-826. DOI: 10.1109/JQE.2012.2189199

[61] Lin L, Wang Q. Microplasma: A new generation of technology for functional nanomaterial synthesis. Plasma Chemistry and Plasma Processing.

2015;**35**(6):925-962. DOI: 10.1007/s11090-015-9640-y

[62] Iza F, Kim GJ, Lee SM, Lee JK, Walsh JL, Zhang YT, et al. Microplasmas: Sources, particle kinetics, and biomedical applications. Plasma Processes and Polymers. 2008;**5**(4): 322-344. DOI: 10.1002/ppap.200700162

[63] Becker KH. The use of nonthermal plasmas in environmental applications. In: Introduction to Complex Plasmas. Berlin Heidelberg: Springer-Verlag; 2010. pp. 367-394. DOI: 10.1007/978-3-642-10592-0

[64] Zhu Z, Chan GC-Y, Ray SJ, Zhang X, Hieftje GM. Microplasma source based on a dielectric barrier discharge for the determination of mercury by atomic emission spectrometry. Analytical Chemistry. 2008;**80**(22):8622-8627. DOI: 10.1021/ac801531j

[65] Joffrion JB, Mills D, Clower W, Wilson CG. On-chip microplasmas for the detection of radioactive cesium contamination in seawater. Micromachines. 2017;**8**(9):259. DOI: 10.3390/mi8090259

[66] Baranov OO, Xu S, Xu L, Huang S, Lim J, Cvelbar U, et al. Miniaturized plasma sources: Can technological solutions help electric micropropulsion? IEEE Transactions on Plasma Science. 2018;**46**(2):230-238. DOI: 10.1109/TPS.2017.2773073

[67] Elkholy A, Shoshyn Y, Nijdam S, van Oijen J, van Veldhuizen E, Ebert U, et al. Burning velocity measurement of lean methane-air flames in a new nanosecond DBD microplasma burner platform. Experimental Thermal and Fluid Science. 2018;**95**:18-26. DOI: 10.1016/j.expthermflusci.2018.01.011

Section 3

Nanogenerators

Chapter 4

Nanogenerators from Electrical Discharge

Jie Wang, Di Liu, Linglin Zhou and Zhong Lin Wang

Abstract

Electrical discharge is generally considered as a negative effect in the electronic industry and often causes electrostatic discharge (ESD) and thus failure of electronic components and integrated circuits (IC). However, this effect was recently used to develop a new energy-harvesting technology, direct-current triboelectric nanogenerator (DC-TENG). In this chapter, its fundamental mechanism and the working modes of the nanogenerator will be presented. They are different from the general alternating current TENG (AC-TENG) invented in 2012, which is based on triboelectrification and electrostatic induction. Taking advantage of the electrostatic discharge, it can not only promote the miniaturization trend of TENG and self-powered systems, but also provide a paradigm shifting technique to in situ gain electrical energy.

Keywords: electrostatic discharge, mechanical energy harvesting, nanogenerators, self-powered systems

1. Introduction

Static electricity is a documented phenomenon since the ancient Greek era of 2600 years ago [1–3]. At that time, people found that the amber through friction can attract lightweight particles, attracting a lot of researchers to study the physical principle behind this interesting phenomenon. Triboelectrification (or contact electrification), which refers to the charge transfer between two surfaces in contact, is the principle behind natural phenomena such as the amber effect and lighting [4–8]. Generally, two different materials will have net negative and positive charges, respectively, after contact or by friction based on their capability of gaining and losing electrons. The material which has strong capability of losing electrons will easily be positively charged, and the other has the tendency to be negatively charged. The presence of triboelectric charges on the surface of dielectric will build a strong electric field. When two materials of different polarities are close to each other, a charged material close to either a metal or ground, the electrostatic field between the two materials may break down the air and finally form electrical discharge.

Electrostatic discharge is a ubiquitous phenomenon in our daily life which is a sudden flow of electricity between two charged materials caused by contact or dielectric breakdown [9–13]. For instance, during dry winters, the body easily accumulates static electricity, and it is very prone to discharge when it makes contact with conductors or other people. Generally, electrical discharge is considered as a negative effect in the electronic industry and often causes electrostatic discharge (ESD) failure of electronic components and integrated circuits (IC). Intensive work has been dedicated to avoiding ESD to protect electronic components and instruments' safety [14–20].

IntechOpen

With the rapid development of science and technology, the structure of energy demand has changed dramatically. Conventional ordered energy cannot fully meet modern society's demand of a clean and sustainable power source with the increasing demand of wearable electronics and Internet of Things (IoTs). We need "disordered energy" to meet the remaining energy demand of electronics, which are widely distributed, possibly moved and large quantity [21]. At present, most electronic devices are powered by batteries and/or local power generators. However, a battery has limited life cycle which has to be constantly monitored, recharged, or replaced, and it needs a lot of manpower and material resources, thus increasing the maintenance cost [22]. With considering the working status of each electronics, a variety of energy-harvesting methods show their respective characteristics. For example, solar cells can harvest solar energy except in the day time when there is sun light [23]; wind power generation [24] works under abundant wind energy; a thermal generator can convert temperature difference into electricity [25]; and piezoelectric nanogenerators can convert tiny physical deformations into electricity to self-power small-scale devices [26]. Based on the triboelectrification effect and electrostatic induction, the use of triboelectric nanogenerators (TENGs) has been demonstrated as a cost-effective, clean, and sustainable strategy to convert mechanical energy into electricity with comprehensive advantages of light weight, small size, a wide choice of materials, and high efficiency even at low frequencies [27–30].

2. Triboelectric nanogenerators

In 2012, triboelectric nanogenerators (TENGs) based on triboelectrification effect and electrostatic induction were invented by Zhong Lin Wang to harvest mechanical energy from ambient environment [31]. In addition, the self-powered systems based on TENGs demonstrated an effective solution to supply energy for micro/nano electronics. A conventional TENG can generate AC by the friction of two materials with different electron affinity, where charge transfer occurs between the two surfaces of materials, and then inducing electron transfer between two back electrodes under the periodical mechanical force. Recent research indicates that its fundamental theory lies in Maxwell's displacement current and change in surface polarization [32]. Based on this principle, four different modes of TENGs are built according to different device structures and working environments: vertical contact-separation mode, lateral siding mode, single-electrode mode, and freestanding triboelectric-layer mode (**Figure 1**) [33]. Based on the four modes of TENG, several works have demonstrated that TENGs can harvest various forms of mechanical energy, such as human motion, vibration, wind and even blue energy, making possible their applications in wearable electronics, remote and mobile environmental sensors, and IoTs [34–38].

As an energy harvester, the output power density is one of the key properties to measure the output capability of TENGs. Recent progress indicates that the power density is quadratically related to triboelectric charge density [39, 40], and thus, great efforts have been concentrated on increasing the triboelectric charge density by means of material improvement, structural optimization, surface modification, and so on [41–43]. Jie Wang et al. fabricated a flexible TENG with the silicon rubber as a triboelectric layer and a mixture of silicon rubber [44], carbon black, and carbon nanotubes as a triboelectric electrode. With optimized structural design, the triboelectric charge density is increased up to 250 $\mu C\ m^{-2}$. By designing a three-layer TENG, the triboelectric charge density increases to ~270 $\mu C\ m^{-2}$, which is the theoretical limit of air breakdown [45]. Then, a high-vacuum environment was adopted to suppress electrostatic breakdown and a triboelectric

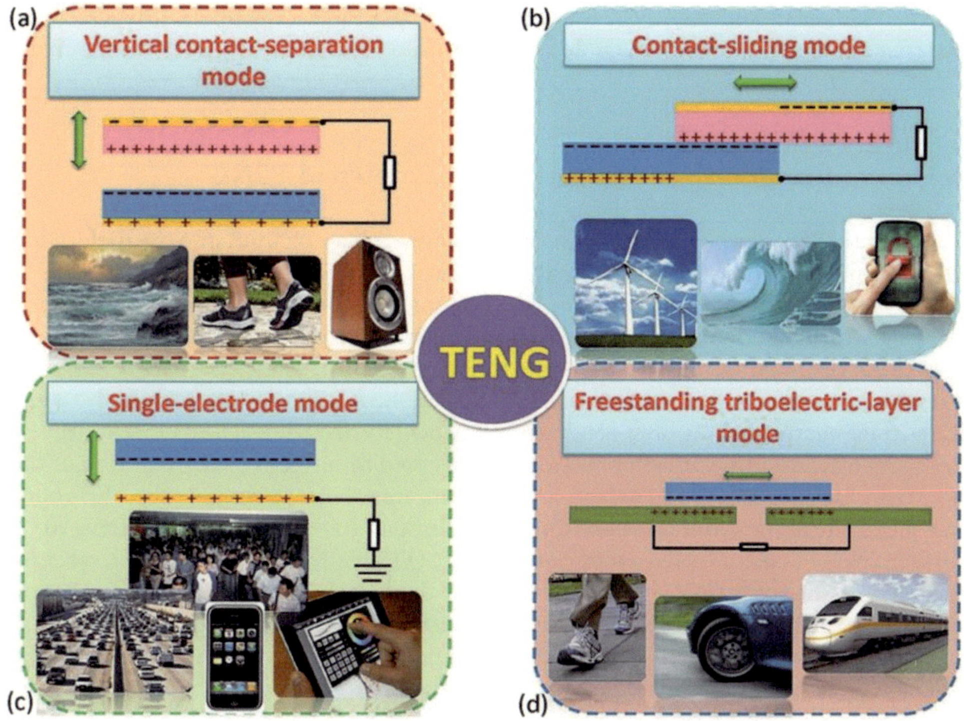

Figure 1.
The four fundamental modes of TENGs: (a) vertical contact-separation mode, (b) in-plane contact-sliding mode, (c) single-electrode mode, and (d) freestanding triboelectric-layer mode [33].

charge density of 660 μC m^{-2} is achieved. By further coupling surface polarization from triboelectrification and hysteretic dielectric polarization from ferroelectric material in vacuum, the triboelectric charge density boosts to 1003 μC m^{-2} without the constraint of air breakdown [46].

Triboelectric charge density as one of the main optimization directions of TENGs is gradually increased from 50 to ~1000 μC m^{-2}, and electrostatic breakdown becomes a problem that must to be considered. Generally, a high electrostatic field will be built between the two charged surfaces with opposite triboelectric charges during the working process of TENG. Paschen's law describes the empirical relationship between gaseous breakdown voltage (V_b), gap distance (x), and gas pressure (P) and is given by

$$V_b = \frac{APd}{ln\,(Pd) + B} \tag{1}$$

where A and B are constants determined by the composition and the pressure of the gas. For air at standard atmospheric pressure (atm, i.e., the conventional operation condition of a TENG), A = 273.75 V Pa^{-1} m^{-1} and B = 1.08.

According to the theoretical derivation, the gap voltage between contact surfaces of a TENG (V_{gap}) under short-circuit condition is given by

$$V_{gap} = \frac{\sigma t d}{\varepsilon_0(t + d\,\varepsilon_r)} \tag{2}$$

where t is the thickness of the polytetrafluoroethylene (PTFE) film, σ the triboelectric surface charge density, ε_r the relative permittivity of PTFE (ε_r ~ 2.1), and ε_0 the vacuum permittivity (ε_0 ~ 8.85 × 10^{-12} F m^{-1}).

To avoid air breakdown, V_{gap} must be smaller than V_b at any operation gap distance. Large efforts have been dedicated to study the electrostatic breakdown of TENGs.

3. The confirmation and study of air breakdown in TENG

Using the ion-injection method for introducing surface charges into the dielectric layer, the power density of TENG greatly increases to ~315 W m^{-2} [39]. With the help of this method, the maximum surface charge density of TENG with the limitation of electrostatic breakdown was observed and confirmed for the first time. As shown in **Figure 2a**, the maximum surface charge density gradually increased with the ion-injection process (the thickness of the utilized FEP film is 50 μm). In the initial state, the surface charge density only generated by triboelectrification is nearly 50 μC m^{-2}. Subsequently, step-by-step ion injection was adopted for effective accumulation of the negative charges on the FEP surface, and it is very important to connect the FEP's bottom electrode to the ground in each ion-injection step. After a few ion injections, the surface charge density increased to ~240 μC m^{-2} (**Figure 2b**). After the ninth injection, the performance of transferred charges became distinctively different that this abrupt decrease of surface charge resulted from air breakdown. The schematic and numerical simulation results showing the voltage drop in the air gap between the Al and FEP layers, which could cause the air breakdown is shown in **Figure 2c**. This provides a new optimization direction for realizing high output performance of TENG.

After confirming the existence of air breakdown, many researchers studied the electrostatic breakdown of TENGs and various experiments demonstrated the

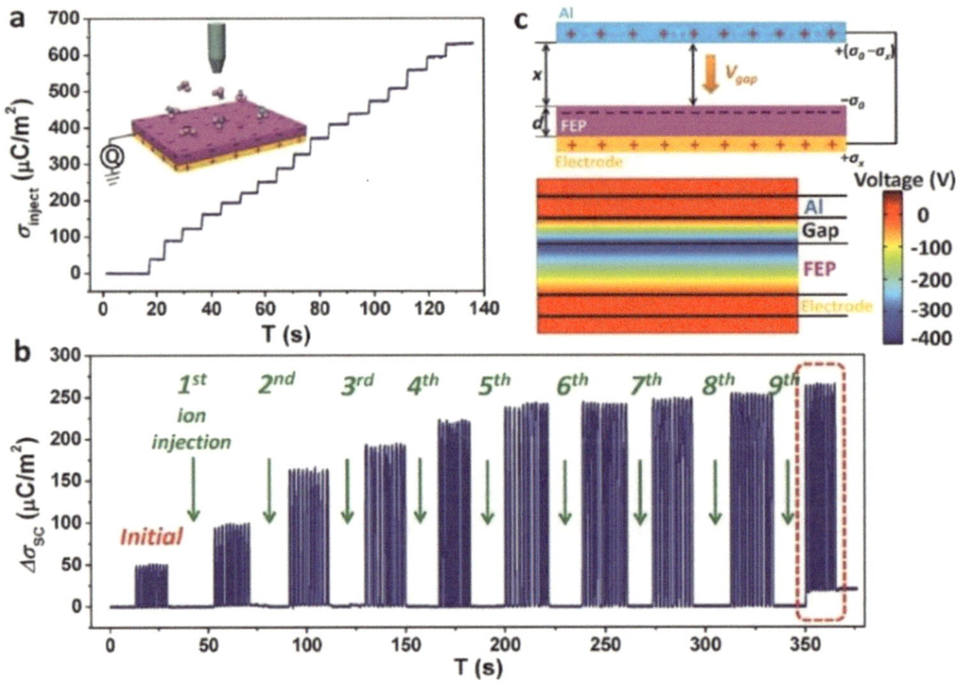

Figure 2.
Step-by-step measurement of the surface charge density by ion-injection process. (a) In situ measurement of the charge density of the FEP film during the step-by-step ion-injection process. (b) The short-circuit charge density ($\Delta\sigma_{SC}$) measured by the TENG when the FEP film was injected with ions time-by-time. (c) Schematic and numerical simulation results showing the voltage drop (V_{gap}) between the Al and the FEP layers, at which the air breakdown may occur [39].

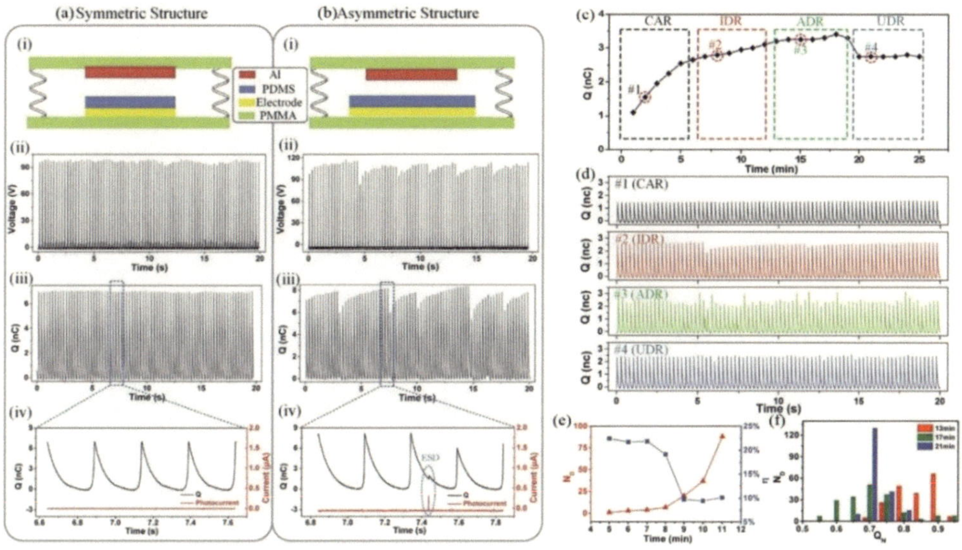

Figure 3.
Comparison of S-TENG with A-TENG, and the demonstration of ESD. (a) The structure (i), stable output voltage (ii), stable Q waveforms (iii), and signals detected by a photocurrent detector of S-TENG (iv). (b) The structure (i), abrupt voltage decline (ii), abrupt declined Q waveforms (iii), and signals detected by a photocurrent detector of A-TENG (iv). (c) Triboelectric charges in four different regions of ESD processes. (d) Picked-up, 20-s waveforms for each marked point in (c). (e) ND and η change with time in IDR. (f) Double-layer feature in three different minutes of ADR [47].

existence of air breakdown. Haixia Zhang et al. [47] fabricated a symmetric structure and an asymmetric structure of TENG, discovered the phenomenon of electrostatic breakdown in the asymmetric structure, and then studied the four steps of electrostatic breakdown in detail. **Figure 3a, b** shows the comparison of the symmetric structure and the asymmetric structure. A low-dark-current photoelectric detector is used to test photocurrent signals and an obvious light signal can be observed between the abrupt output decline in **Figure 3b(iv)**, while no light signal has been detected in **Figure 3a(iv)**. The transition process of electrostatic breakdown has been investigated and summarized, and the changes in triboelectric charges with time at four different regions in electrostatic breakdown, which are charge accumulation region (CAR), intermittent discharge region (IDR), accelerated discharge region (ADR), and uniform discharge region (UDR) are shown in **Figure 3c–f**. Meanwhile, the influence of several factors including contact materials, contact pressure, titled angle, and surface morphology on electrostatic breakdown has been studied.

Yunlong Zi et al. [48] confirmed that the threshold surface charge density of contact-separation mode TENG is nearly 40~50 μC m^{-2} with the existence of air breakdown by the theoretical study of the maximized effective energy output, and then the effects of gas pressure (higher than the atmosphere) and gas composition were also studied. **Figure 4a** shows the breakdown voltage as calculated by Paschen's law in 1 atm air, in which the points A–E show the voltage V_1 between triboelectric surfaces of contact-separation mode TENG under different transferred charges. The potential distribution of A–D is simulated by COMSOL Multiphysics software shown in **Figure 4b**. In **Figure 4c,d** are shown the voltage V_1 between the two triboelectric surfaces and the breakdown voltage V_b. To further clearly explain the relationship of V_1 and V_b, **Figure 4e** shows the distribution of V_b-V_1 in a V-Q plot with the contour line of 0 V displayed as the red dashed line. The negative ("–") and positive ("+") areas are divided by this contour line. They also did experiments to demonstrate the existence of air breakdown. **Figure 4f** is the mechanism of contact-separation mode TENG when the air breakdown existed. The final charge densities of six TENGs with different initial charge densities are shown in **Figure 4g**. The decrease in final charge densities of TENG #3–6 indicates the existence of air breakdown.

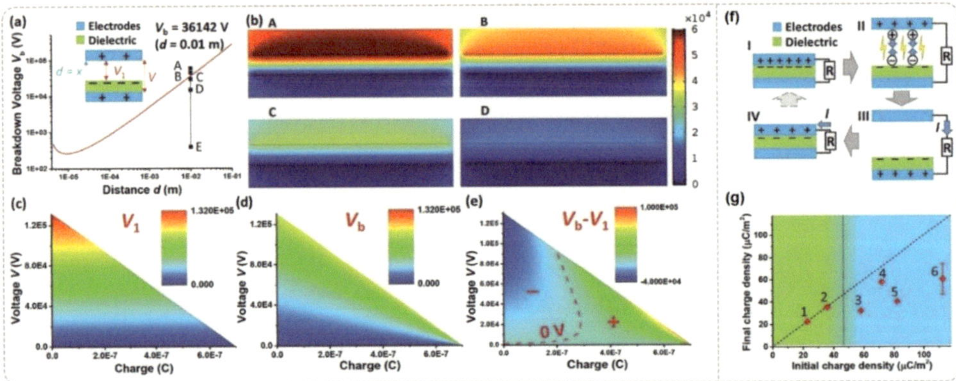

Figure 4.
The demonstration of air breakdown in a CS mode TENG. (a) The breakdown voltage calculated by Paschen's law in 1 atm air, in which the points A–E show the voltage V_1 between dielectric layer and upper electrode of the CS mode TENG with different surface charge density (inset shows the schematic diagram of the TENG). (b) The potential distribution of A–D simulated by COMSOL Multiphysics software. (c) The voltage V_1 between the dielectric layer and upper electrode in V-Q plot. (d) The breakdown voltage V_b in V-Q plot. (e) The distribution of V_b-V_1 in V-Q plot. The red dashed line is the contour line of 0 V. (f) The working process of CS mode TENG with air breakdown and the mechanism of the final charge density measurement. (g) The test results of final charge densities for six TENGs. The equal initial and final charge densities are shown with the dashed inclined line, and the dotted vertical line indicates the existence of air breakdown where σ_t separates with σ_o (blue) [48].

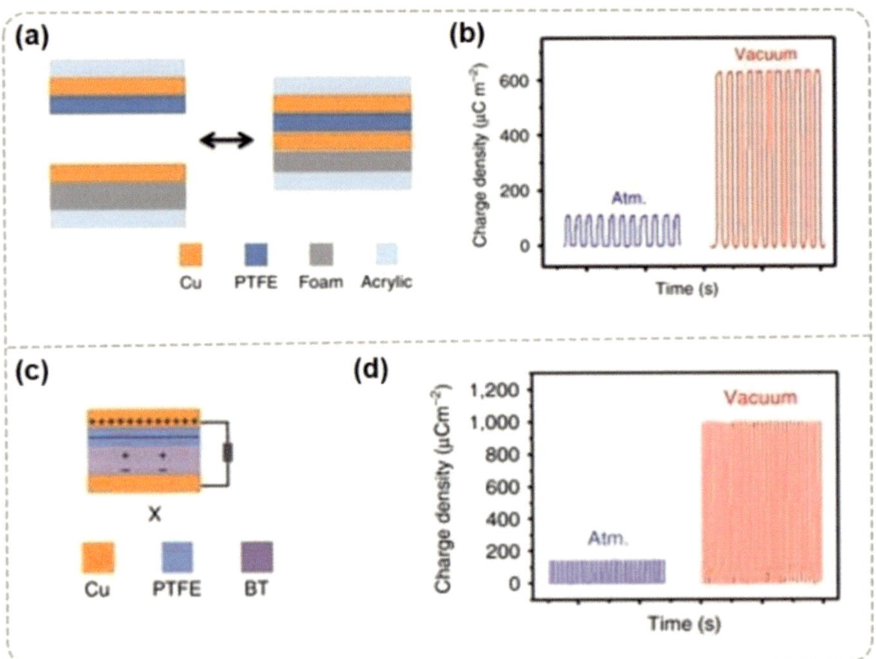

Figure 5.
Output performance of TENG in vacuum. (a) Schematic of TENG with a cushioned Cu electrode to increase contact intimacy during operation process. (b) Charge density of the TENG in atmosphere and high vacuum ($P{\sim}10^{-6}$ torr). (c) Schematic of the TENG with the integration of triboelectric material (PTFE) with ferroelectric material (BT). (d) Charge density of the TENG in atmosphere and high vacuum [46].

Besides the relatively theoretical research of air breakdown, great efforts have been devoted to suppress or avoid air breakdown for substantially enhancing the output performance of TENGs [49, 50]. On the basis of soft contact and fragmental structure, Jie Wang et al. [46] fabricated a TENG with a cushioned Cu electrode and a contact area smaller than conventional TENG to improve the contact intimacy, which is shown in **Figure 5a**. After working in high-vacuum environment (~10^{-6} Torr), the triboelectric charge density is increased up to 660 μC m^{-2} (**Figure 5b**). Meanwhile, by properly combining the surface polarization from triboelectrification and residual dielectric polarization of ferroelectric materials, the surface charge density can be expected to further increase. Then, a TENG with the integration of triboelectric material PTFE with ferroelectric material BT (one type of doped barium titanate material) in vacuum is fabricated (**Figure 5c**), and its surface charge density jumped to 1003 μC m^{-2} without the constraint of air breakdown (**Figure 5d**). The surface polarization from triboelectrification induces the dielectric polarization in BT, and the latter further enhances the former until a new equilibrium rebuilds again. This combination of surface polarization and dielectric polarization greatly enhances the triboelectric charge density under the rest cycles of TENG. This work proposes a new optimized method to realize high-output power density TENG and meanwhile avoids the effect of environmental factors on the output performance of TENG.

4. Applications of air breakdown in conventional TENGs

TENGs can convert all kinds of mechanical energy into electricity, and have a built-in characteristic, that is high output voltage. Taking advantage of the high-output voltage characteristic, a lot of practical applications of TENG have been

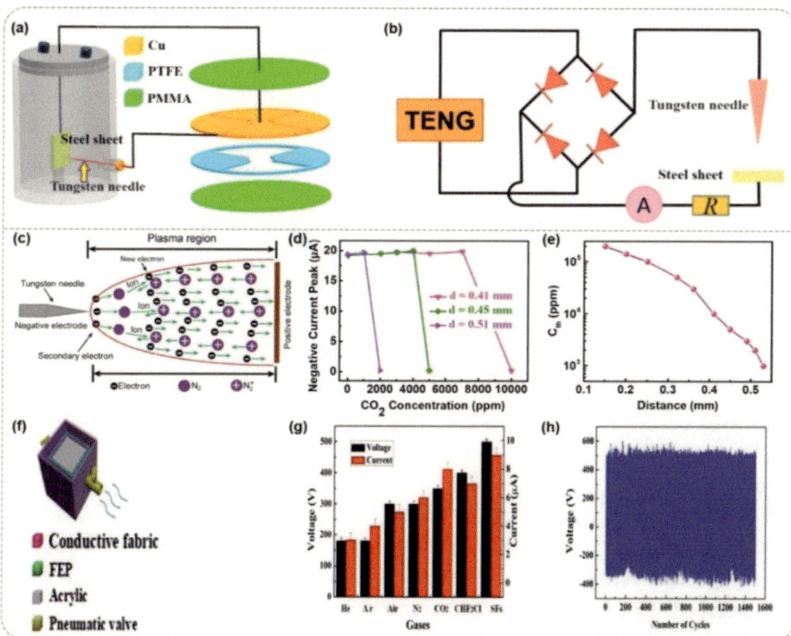

Figure 6.
Self-powered gas sensors. (a) Structure diagram of the self-powered CO_2 sensor. (b) Circuit diagram of the self-powered CO_2 sensor. (c) Schematic image of the gas discharge process in pure N_2. (d) Under negative voltage, the curves of discharge current peak at different "d" under different CO_2 concentrations. (e) The curve of C_{th} at different "d" [49]. (f) Schematic diagram of the cubic-TENG structure. (g) Output performance under various gas atmospheres of He, Ar, air, N_2, CO_2, CHF_2Cl, and SF_6. (h) Long-term stability under an SF_6 atmosphere [55].

developed and successfully demonstrated in generating the input ions in mass spectrometry [51], fabricating electrospun nanofibers [52], driving field emission of electrons [53] etc. Considering the effects of gas composition and gas pressure on air breakdown, great efforts have been dedicated to study the air discharge for gas sensors. Ke Zhao et al. [54] fabricated a self-powered CO_2 gas sensor based on gas discharge induced by a TENG. The structure diagram of a self-powered CO_2 gas sensor is shown in **Figure 6a**. **Figure 6b** is the circuit diagram of the self-powered CO_2 gas sensor. The detailed gas discharge process is shown in **Figure 6c**. N_2 molecules lose electrons under the bombardment of electrons to form positive N_2 ions. N_2 molecules and electrons are accelerated to different directions under the electric field. Since the mass of positive N_2 ions is much higher than that of the electrons, electrons are easier to be accelerated. The accelerated electrons will bombard other N_2 molecules to form new positive ions and electrons; eventually an electron avalanche is formed. It is shown that the C_{th} (C_{th} is defined as the critical CO_2 concentration that causes the stop of gas discharge) decreased with the increase of distance. Shasha Lv et al. [55] fabricated an enclosed cubic-TENG consisting of an inner and outer box using a fluorinated ethylene propylene thin film and a layer of conducting fabric as the triboelectric layers, which is shown in **Figure 6f**. The output performance of the cubic-TENG in different gas atmospheres is shown in **Figure 6g**. Long-term stability under SF6 atmosphere is shown in **Figure 6h,** indicating the potential of cubic-TENG as an energy-harvesting device and vibration sensor.

5. Mechanical energy harvesting via air breakdown

5.1 DC-TENG

Besides the applications of air breakdown in TENGs for gas sensors, some works about using air breakdown producing electricity also have been studied. Ya Yang et al. [56] fabricated a DC triboelectric generator (DC-TEG) consisting of two wheels and a belt, which is shown in **Figure 7a**. **Figure 7b** is a scanning electron microscope (SEM) image of the surface of PTFE. The working mechanism of the DC-TEG is illustrated in **Figure 7c** (The reference point T was used to monitor the relative sliding length of the belt). The design of the DC-TEG relied on the different properties to lose electrons, which is $\alpha_{III} > \alpha_{I} > \alpha_{II}$ in this structure. Here, the three materials are an Al wheel, a rubber belt, and a PTFE wheel. When the belt I makes contact with the wheel III, electrons will transfer from wheel III to belt I, resulting in net negative charges on the inner surface of belt I. After belt I makes contact with wheel II, electrons will transfer from belt I to wheel II. With the rotation of two wheels, more and more positive and negative charges will accumulate on the surface of wheel III and wheel II. Moreover, some positive charges are accumulated at electrode 2 (E2) due to electrostatic induction. So, the high electric field between wheel II and E2 will breakdown the air during the gap forming a pulsed DC output in external circuit. This DC-TEG exhibited good performance with different rotation rates and was demonstrated to power electronic systems directly. **Figure 7d** shows 1020 LEDs in series to fabricate LED panels. All the LEDs can be driven by the DC-TEG at a rotation speed of 3044 r min^{-1} as shown in **Figure 7e**. **Figure 7f** depicts charging curves of a 1-μF capacitor charged by the DC-TEG at different rotation rates. The constant-current discharging curves of a 1000-μF capacitor after being charged by DC-TEG is shown in **Figure 7g**, where the discharging current is 60 μA.

Jianjun Luo et al. [57] reported a DC-TENG realized by air breakdown-induced ionized air channel. The structure of the DC-TENG is illustrated in **Figure 8a**. This DC-TENG is composed of a top triboelectric aluminum (Al) electrode (noted as

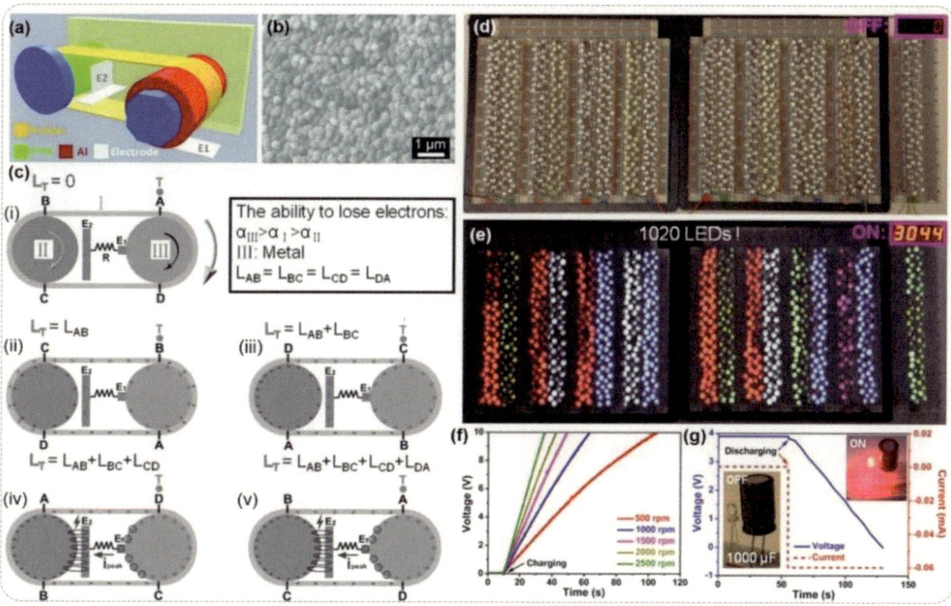

Figure 7.
DC-TEG. (a) Schematic diagram of the DC-TEG. (b) SEM image of the PTFE surface. (c) Working mechanism of the DC-TEG. (i) Initial status without sliding motion of the belt. (ii–v) The triboelectric charge distributions when belt I went through the point T with the length of L_{AB}, $L_{AB} + L_{BC}$, $L_{AB} + L_{BC} + L_{CD}$, and $L_{AB} + L_{BC} + L_{CD} + L_{DA}$, respectively. L_T is the length that the belt went through the reference point T. L_{AB} is the length of the belt between A and B. (d), (e) Photographs of 1020 LEDs driven by the DC-TEG as a direct power source when the DC-TEG is (d) off and (e) on at a rotational speed of 3044 r min^{-1}. (f) The measured voltage of the DC-TEG charges a 1-μF capacitor at different rotational speeds. (g) The constant-current discharging curves of a 1000-μF capacitor after charging by the DC-TEG (inset shows the photographs of a red LED powered by the charged capacitor) [56].

Al-I), a sponge as a compression layer, a FEP film adhered to a back Al electrode (noted as Al-II), and an Al electrode at the bottom right corner (noted as Al-III). **Figure 8b** is the photograph of the DC-TENG. **Figure 8c,d** shows the transferred charges and short-circuit current of the DC-TENG with DC output characteristic. Using polydimethylsiloxane (PDMS), Kapton, and polyethylene terephthalate (PET) as triboelectric layers to replace FEP, the output characteristic is not changed, indicating the DC output is universal (**Figure 8e**). The working mechanism of the DC-TENG is illustrated in **Figure 8f**. The charge transfer between the top electrode and the dielectric layer is based on triboelectrification. In a working process, the electrons transport from bottom electrode to top electrode relying on an external circuit, which is Al-III in this structure. Then, the electrons flow back from top electrode to bottom electrode via the ionized air channel created by air breakdown. Because the inner flow of electrons based on air breakdown is in a single direction, the output current will be pulsed DC. This working mechanism was verified by real-time electrode potential monitoring, photocurrent signal detection, and controllable discharging observation. A flexible DC-TENG also fabricated to demonstrate this device can drive electronics directly without a rectifier, and the circuit diagram is illustrated in **Figure 8g**. **Figure 8h** shows the voltage curves of a 1-μF capacitor charged by the DC-TENG at different frequencies. The DC-TENG can also integrate with a capacitor and a calculator to form a self-powered system (**Figure 8i**).

5.2 Constant-current TENG arising from electrostatic breakdown

TENGs are considered as a potential solution via building self-powered systems. The conventional TENGs have two built-in characteristics (i.e., AC consisted of

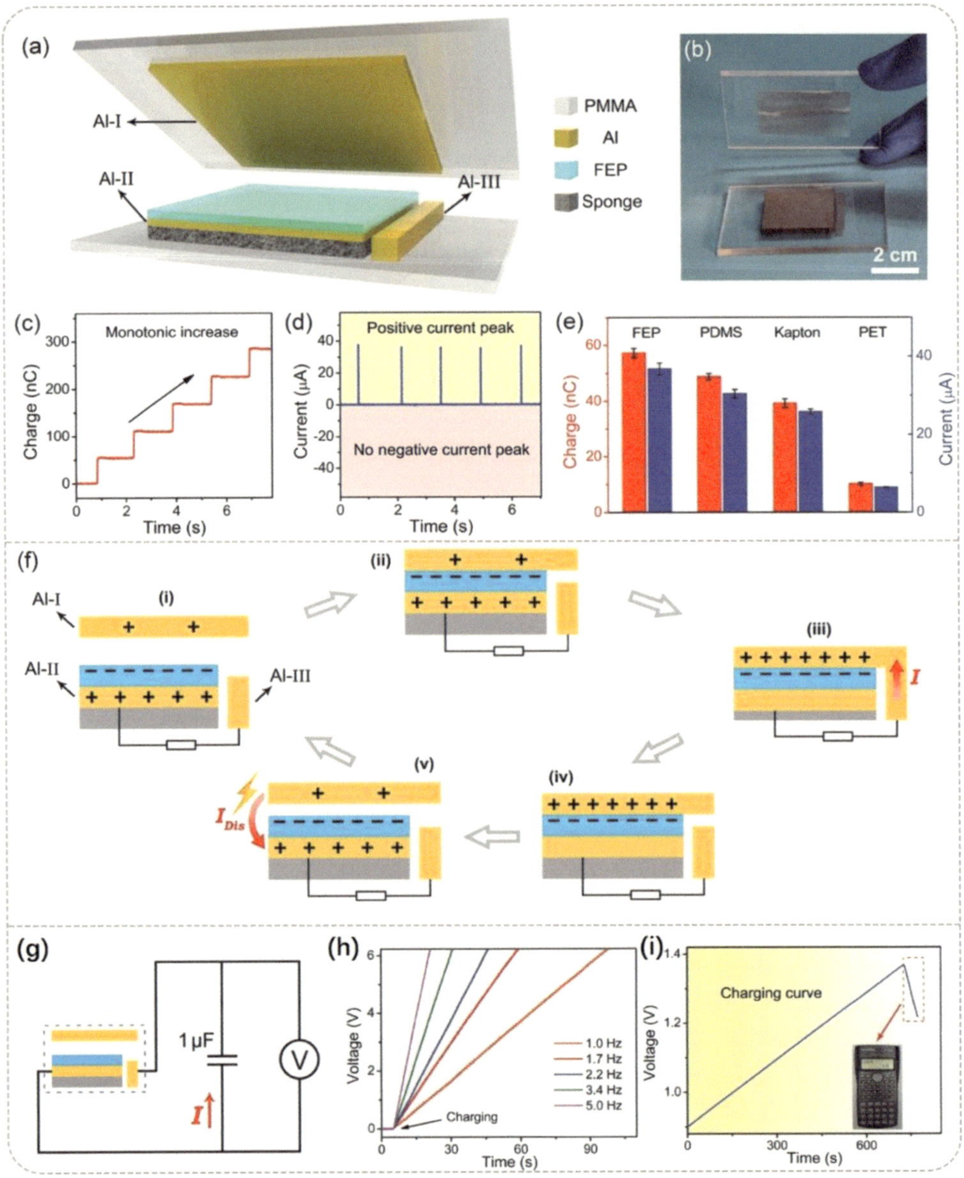

Figure 8.
DC-TENG. (a) Structure of the DC-TENG. (b) Photograph of the DC-TENG. (c) Transferred charges of the DC-TENG. (d) Short-circuit current of the DC-TENG. (e) Output performance of DC-TENG using different triboelectric materials. (f) Schematic working principle of the DC-TENG achieved through the ionized air channel caused by air breakdown. (g) Circuit diagram of a 1-μF capacitor directly and continuously charged by the DC-TENG without a rectifier. (h) Measured voltage of a 1-μF capacitor charged by the DC-TENG at different frequencies. (i) Charging curve of a 1000-μF capacitor charged by the flexible multilayered DC-TENG (inset shows the photograph of a calculator powered by the charged capacitor). (i-v) The working mechanism of the DC-TENG in stable working state [57].

pulse series). Generally, TENGs need to connect with a rectifier or a capacitor to drive electronics, which takes away its portability advantage [58, 59]. Second, the pulsed output current of TENGs result in high crest factor, which is a key metric to output instability defined as the ratio of the peak value to the root mean square value. This greatly influences their performance for energy storage efficiency and powering electronics [59]. In addition, triboelectric charge density as one of the key properties of TENGs has been greatly increased; therefore, a very high

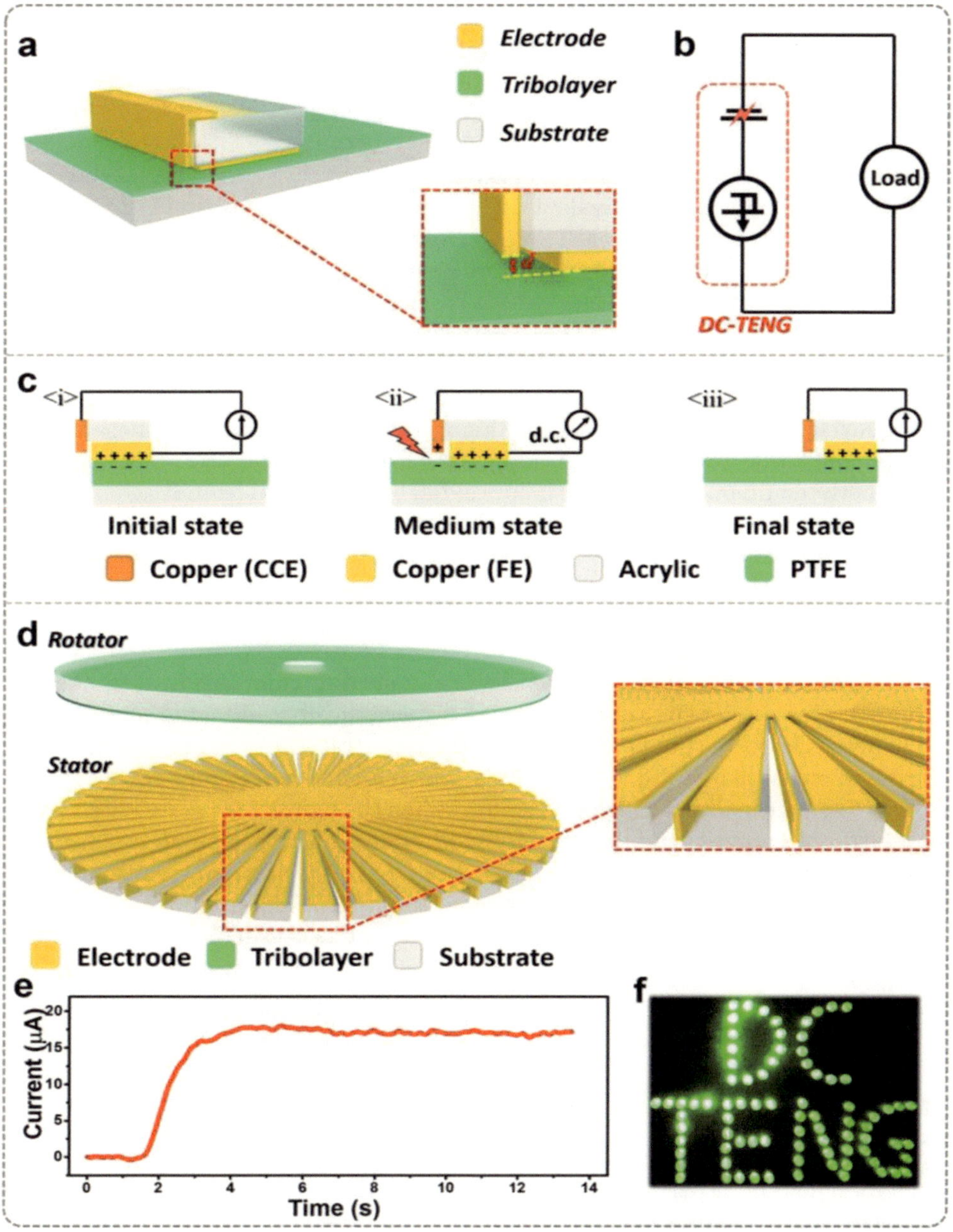

Figure 9.
Constant-current TENG. (a) A schematic illustration of the sliding mode DC-TENG. (b) Equivalent circuit model of the DC-TENG. (c) Working mechanism of the sliding mode DC-TENG in full cyclic motion. (d) Structural design of the rotary mode DC-TENG. Inset shows a zoomed-in illustration of its stator. (e) Constant current output of the DC-TENG. (f) Photograph of 81 LEDs with stable luminance powered by a rotary mode DC-TENG [60].

electrostatic field will be built in TENG, which also leads to air breakdown and results in unwanted charge quantity loss and, consequently, quadratic loss in output power. The charge quantity loss can be roughly estimated with its charge density gap in air and vacuum. With a 50-μm PTFE film as a triboelectric layer, 240 $\mu C\ m^{-2}$ is the theoretical upper limit in air [39, 44], but 1003 $\mu C\ m^{-2}$ has been achieved in vacuum, where a dominant part is wasted because of air breakdown [46].

To harvest energy during electrostatic breakdown, Jie Wang et al. [60] designed a next-generation DC-TENG via the triboelectrification effect and electrostatic breakdown, which consists of a frictional electrode (FE), a charge collecting electrode (CCE), and a triboelectric layer (**Figure 9a**). The CCE layer is fixed on the side of a sliding acrylic substrate, with a subtle distance to the triboelectric layer, which is a PTFE layer attached to another acrylic sheet. The CCE and FE both are copper electrodes here. The physics model of the new DC-TENG is made up of an electric charge source and a broken-down capacitor composed of the CCE and PTFE film, as its equivalent circuit briefly demonstrates in **Figure 9b**. It is different with a conventional TENG, whose paradigm is a variable capacitor initially charged by triboelectrification and generating AC pluses by electrostatic induction. The working mechanism of the DC-TENG is shown in **Figure 9c**. When the FE makes contact with the PTFE, electrons will transfer from the FE to PTFE based on triboelectrification effect. PTFE as an electret can hold a quasi-permanent electric charge (**Figure 9c, i**). Thus, when the slider moves forward, the electrons on the surface of PTFE will build a very high electric field between the negatively charged PTFE film and the CCE. As long as it exceeds the dielectric strength of the air between them, whose value is approximately 3 kV/mm from Paschen's law, it can cause the nearby air to partially ionize and begin conducting. Electrons will transfer from PTFE film to CCE (**Figure 9c, ii**); that is, the CCE is rationally placed to induce air breakdown, creating artificial lightning. When the slider is stationary on the surface of PTFE, electrons will stop transfer (**Figure 9c, iii**). Because the inner flow direction of electrons is fixed from the FE to the PTFE film and then to the CCE, the output electrons will also be in a single direction, that is, from the CCE to the FE. Thus, cyclic DC can be produced by periodically sliding the slider.

To optimize the output performance, a radially arrayed rotary of DC-TENG is fabricated by parallel multiple DC-TENGs, which is shown in **Figure 9d** and inset shows a zoomed-in illustration of its stator. When the motor rotates steadily, the crest factor of output current is very close to 1, indicating approximately constant current output characteristic (**Figure 9e**). This DC-TENG was demonstrated to charge capacitors or drive electronics without a rectifier. A light-emitting diode (LED) bulb arrays can also be lit up by the rotary mode DC-TENG with a rotation rate of 500 r min^{-1} (**Figure 9f**). Unlike when driven by the conventional AC-TENG, the LEDs remain at constant luminance without flashing lights.

6. Conclusions

Harvesting of environmental mechanical energy as an eco-friendly energy generation method is particularly a promising solution and plays an increasingly important role in driving wearable electronics and sensor networks in the IoTs. Based on the triboelectrification effect and electrostatic induction, the use of TENG invented in 2012 by Zhong Lin Wang has been demonstrated as a cost-effective, clean, and sustainable strategy to convert mechanical energy into electricity with comprehensive advantages of light-weight, small size, a wide choice of materials, and high efficiency even at low frequencies, which shows great potential in promoting the miniaturization trend of self-powered systems. With the gradually increase in triboelectric charge density, electrostatic breakdown, which is generally considered as a negative effect in the conventional TENG, becomes an issue that must to be considered. The theoretical and experimental studies of air breakdown in TENGs are important to promote the development of this field. By taking advantage of the electrostatic discharge, several types of DC-TENG have been demonstrated to power electronics directly without a rectifier or a capacitor. Comparing with the output characteristics of conventional TENG (AC consisted of pulse series), a constant-current output (crest factor ~1) is achieved by coupling triboelectrification effect and electrostatic breakdown. This constant-current TENG has been demonstrated to drive LED bulb arrays that remain at constant luminance without flashing lights. Moreover, it can not only promote the miniaturization trend of TENG and self-powered systems but also provide a paradigm shifting technique to in situ gain electrical energy. Based on the above discussion and analysis, the electrostatic breakdown in TENG will soon become a hot issue and we would require new studies in electrostatic breakdown.

Acknowledgements

The authors would like to thank the financial supports from the National Key R& D Project from Minister of Science and Technology (2016YFA0202704) and National Natural Science Foundation of China (Grant Nos. 61774016, 21773009, 51432005, 5151101243, 51561145021).

Conflict of interest

There is no conflict of interest.

Author details

Jie Wang[1,2]*, Di Liu[1,2], Linglin Zhou[1,2] and Zhong Lin Wang[1,2,3]

1 Beijing Institute of Nanoenergy and Nanosystems, Chinese Academy of Sciences, Beijing, P. R. China

2 College of Nanoscience and Technology, University of Chinese Academy of Sciences, Beijing, P. R. China

3 School of Materials Science and Engineering, Georgia Institute of Technology, Atlanta, GA, USA

*Address all correspondence to: wangjie@binn.cas.cn

References

[1] Ballou JW. Static electricity in textiles. Textile Research Journal. 1954;**24**(2):146-155. DOI: 10.1177/004051755402400209

[2] Shashoua VE. Static electricity in polymers. 1. Theory and measurement. Journal of Polymer Science. 1958;**33**(126):65-85. DOI: 10.1002/pol.1958.1203312608

[3] Shashoua VE. Static electricity in polymers. 2. Chemical structure and antistatic behavior. Journal of Polymer Science Part A: General Papers. 1963;**1**(1):169-187. DOI: 10.1002/pol.1963.100010114

[4] Lowell J, Roseinnes AC. Contact electrification. Advances in Physics. 1980;**29**(6):947-1023. DOI: 10.1080/00018738000101466

[5] Terris BD, Stern JE, Rugar D, Mamin HJ. Contact electrification using force microscopy. Physical Review Letters. 1989;**63**(24):2669-2672. DOI: 10.1103/PhysRevLett.63.2669

[6] Baytekin HT, Patashinski AZ, Branicki M, Baytekin B, Soh S, Grzybowski BA. The mosaic of surface charge in contact electrification. Science. 2011;**333**(6040):308-312. DOI: 10.1126/science.1201512

[7] Horn RG, Smith DT. Contact electrification and adhesion between dissimilar materials. Science. 1992;**256**(5055):362-364. DOI: 10.1126/science.256.5055.362

[8] Lacks DJ, Sankaran RM. Contact electrification of insulating materials. Journal of Physics D: Applied Physics. 2011;**44**(45):453001. DOI: 10.1088/0022-3727/44/45/453001

[9] Talawar MB, Agrawal AP, Anniyappan M, Wani DS, Bansode MK, Gore GM. Primary explosives: Electrostatic discharge initiation, additive effect and its relation to thermal and explosive characteristics. Journal of Hazardous Materials. 2006;**137**(2):1074-1078. DOI: 10.1016/j.jhazmat.2006.03.043

[10] Greason WD. Electrostatic discharge–A charge driven phenomenon (reprinted from electrical overstress electrostatic discharge symposium proceedings, Eos-13, Las-Vegas, Nv, Usa, September 24-26, 1991). Journal of Electrostatics. 1992;**28**(3):199-218. DOI: 10.1016/0304-3886(92)90073-3

[11] Kasemir HW. A contribution to the electrostatic theory of a lightning discharge. Journal of Geophysical Research. 1960;**65**(7):1873-1878. DOI: 10.1029/JZ065i007p01873

[12] Voldman SH. ESD: Physics and Devices[M]. England: John Wiley & Sons; 2005

[13] Voldman SH. Lightning rods for nanoelectronics[J]. Scientific American. 2002;**287**(4):90-97

[14] Vinson JE, Liou JJ. Electrostatic discharge in semiconductor devices: An overview. Proceedings of the IEEE. 1998;**86**(2):399-418. DOI: 10.1109/5.659493

[15] Ker MD, Hsu KC. Overview of on-chip electrostatic discharge protection design with SCR-based devices in CMOS integrated circuits. IEEE Transactions on Device and Materials Reliability. 2005;**5**(2):235-249. DOI: 10.1109/tdmr.2005.846824

[16] Voldman SH. The state of the art of electrostatic discharge protection: Physics, technology, circuits, design, simulation, and scaling. IEEE Journal of Solid-State Circuits. 1999;**34**(9):1272-1282. DOI: 10.1109/4.782088

[17] Wallash A, Kim YK. Magnetic changes in GMR heads caused by electrostatic discharge. IEEE Transactions on Magnetics. 1998;**34**(4):1519-1521. DOI: 10.1109/20.706602

[18] Voldman SH. ESD: Failure Mechanisms and Models[M]. United Kingdom: John Wiley & Sons; 2009

[19] Voldman SH. Electrical Overstress (EOS): Devices, Circuits and Systems[M]. United Kingdom: John Wiley & Sons; 2013

[20] Voldman SH. ESD: Circuits and Devices[M]. England: John Wiley & Sons; 2015

[21] Wang ZL. Entropy theory of distributed energy for internet of things. Nano Energy. 2019;**58**:669-672. DOI: 10.1016/j.nanoen.2019.02.012

[22] Larcher D, Tarascon JM. Towards greener and more sustainable batteries for electrical energy storage. Nature Chemistry. 2015;**7**(1):19-29. DOI: 10.1038/nchem.2085

[23] Law M, Greene LE, Johnson JC, Saykally R, Yang PD. Nanowire dye-sensitized solar cells. Nature Materials. 2005;**4**(6):455-459. DOI: 10.1038/nmat1387

[24] Foley AM, Leahy PG, Marvuglia A, McKeogh EJ. Current methods and advances in forecasting of wind power generation. Renewable Energy. 2012;**37**(1):1-8. DOI: 10.1016/j.renene.2011.05.033

[25] Kraemer D, Poudel B, Feng HP, Caylor JC, Yu B, Yan X, et al. High-performance flat-panel solar thermoelectric generators with high thermal concentration. Nature Materials. 2011;**10**(7):532-538. DOI: 10.1038/nmat3013

[26] Wang ZL, Song JH. Piezoelectric nanogenerators based on zinc oxide nanowire arrays. Science. 2006;**312**(5771):242-246. DOI: 10.1126/science.1124005

[27] Zhu G, Chen J, Zhang TJ, Jing QS, Wang ZL. Radial-arrayed rotary electrification for high performance triboelectric generator. Nature Communications. 2014;**5**:3426. DOI: 10.1038/ncomms4426

[28] Jing QS, Xie YN, Zhu G, Han RPS, Wang ZL. Self-powered thin-film motion vector sensor. Nature Communications. 2015;**6**:8031. DOI: 10.1038/ncomms9031

[29] Wang SH, Lin L, Wang ZL. Triboelectric nanogenerators as self-powered active sensors. Nano Energy. 2015;**11**:436-462. DOI: 10.1016/j.nanoen.2014.10.034

[30] Niu SM, Wang XF, Yi F, Zhou YS, Wang ZL. A universal self-charging system driven by random biomechanical energy for sustainable operation of mobile electronics. Nature Communications. 2015;**6**:8975. DOI: 10.1038/ncomms9975

[31] Fan FR, Tian ZQ, Wang ZL. Flexible triboelectric generator. Nano Energy. 2012;**1**(2):328-334. DOI: 10.1016/j.nanoen.2012.01.004

[32] Wang ZL. On Maxwell's displacement current for energy and sensors: The origin of nanogenerators. Materials Today. 2017;**20**(2):74-82. DOI: 10.1016/j.mattod.2016.12.001

[33] Wang ZL. Triboelectric nanogenerators as new energy technology and self-powered sensors–Principles, problems and perspectives. Faraday Discussions. 2014;**176**:447-458. DOI: 10.1039/C4FD00159A

[34] Zi YL, Lin L, Wang J, Wang SH, Chen J, Fan X, et al. Triboelectric-pyroelectric-piezoelectric hybrid cell for high-efficiency energy-harvesting and self-powered sensing. Advanced Materials. 2015;**27**(14):2340-2347. DOI: 10.1002/adma.201500121

[35] Wen Z, Yeh MH, Guo HY, Wang J, Zi YL, Xu WD, et al. Self-powered textile for wearable electronics by hybridizing fiber-shaped nanogenerators, solar cells, and supercapacitors. Science Advances. 2016;**2**(10):e1600097. DOI: 10.1126/sciadv.1600097

[36] Wang J, Li XH, Zi YL, Wang SH, Li ZL, Zheng L, et al. A flexible fiber-based supercapacitor-triboelectric-nanogenerator power system for wearable electronics. Advanced Materials. 2015;**27**(33):4830-4836. DOI: 10.1002/adma.201501934

[37] Guo HY, Pu XJ, Chen J, Meng Y, Yeh MH, Liu GL, et al. A highly sensitive, self-powered triboelectric auditory sensor for social robotics and hearing aids. Science robotics. 2018;**3**(20):eaat2516. DOI: 10.1126/scirobotics.aat2516

[38] Cheng P, Guo HY, Wen Z, Zhang CL, Yin X, Li XY, et al. Largely enhanced triboelectric nanogenerator for efficient harvesting of water wave energy by soft contacted structure. Nano Energy. 2019;**57**:432-439. DOI: 10.1016/j.nanoen.2018.12.054

[39] Wang SH, Xie YN, Niu SM, Lin L, Liu C, Zhou YS, et al. Maximum surface charge density for triboelectric nanogenerators achieved by ionized-air injection: Methodology and theoretical understanding. Advanced Materials. 2014;**26**(39):6720-6728. DOI: 10.1002/adma.201402491

[40] Zi YL, Niu SM, Wang J, Wen Z, Tang W, Wang ZL. Standards and figure-of-merits for quantifying the performance of triboelectric nanogenerators. Nature Communications. 2015;**6**:8376. DOI: 10.1038/ncomms9376

[41] Wang J, Wen Z, Zi YL, Zhou PF, Lin J, Guo HY, et al. All-plastic-materials based self-charging power system composed of triboelectric nanogenerators and supercapacitors. Advanced Functional Materials. 2016;**26**(7):1070-1076. DOI: 10.1002/adfm.201504675

[42] Yin X, Liu D, Zhou LL, Li XY, Zhang CL, Cheng P, et al. Structure and dimension effects on the performance of layered triboelectric nanogenerators in contact-separation mode. ACS Nano. 2019;**13**(1):698-705. DOI: 10.1021/acsnano.8b07935

[43] Wang SH, Zi YL, Zhou YS, Li SM, Fan FR, Lin L, et al. Molecular surface functionalization to enhance the power output of triboelectric nanogenerators. Journal of Materials Chemistry A. 2016;**4**(10):3728-3734. DOI: 10.1039/c5ta10239a

[44] Wang J, Li SM, Yi F, Zi YL, Lin J, Wang XF, et al. Sustainably powering wearable electronics solely by biomechanical energy. Nature Communications. 2016;**7**:12744. DOI: 10.1038/ncomms12744

[45] Chun JS, Ye BU, Lee JW, Choi D, Kang CY, Kim SW, et al. Boosted output performance of triboelectric nanogenerator via electric double layer effect. Nature Communications. 2016;**7**:12985. DOI: 10.1038/ncomms12985

[46] Wang J, Wu CS, Dai YJ, Zhao ZH, Wang A, Zhang TJ, et al. Achieving ultrahigh triboelectric charge density for efficient energy harvesting. Nature Communications. 2017;**8**:88. DOI: 10.1038/s41467-017-00131-4

[47] Su ZM, Han MD, Cheng XL, Chen HT, Chen XX, Zhang HX. Asymmetrical

triboelectric nanogenerator with controllable direct electrostatic discharge. Advanced Functional Materials. 2016;**26**(30):5524-5533. DOI: 10.1002/adfm.201600909

[48] Zi YL, Wu CS, Ding WB, Wang ZL. Maximized effective energy output of contact-separation triggered triboelectric nanogenerators as limited by air breakdown. Advanced Functional Materials. 2017;**27**(24):1700049. DOI: 10.1002/adfm.201700049

[49] Xu ZS, Duan JJ, Li WB, Wu N, Pan Y, Lin SZ, et al. Boosting the efficient energy output of electret nanogenerators by suppressing air breakdown under ambient conditions. ACS Applied Materials & Interfaces. 2019;**11**(4):3984-3989. DOI: 10.1021/acsami.8b19599

[50] Jiang C, Dai KR, Yi F, Han YZ, Wang XF, You Z. Optimization of triboelectric nanogenerator load characteristics considering the air breakdown effect. Nano Energy. 2018;**53**:706-715. DOI: 10.1016/j.nanoen.2018.09.036

[51] Li AY, Zi YL, Guo HY, Wang ZL, Fernandez FM. Triboelectric nanogenerators for sensitive nano-coulomb molecular mass spectrometry. Nature Nanotechnology. 2017;**12**(5):481-487. DOI: 10.1038/nnano.2017.17

[52] Li CJ, Yin YY, Wang B, Zhou T, Wang JN, Luo JJ, et al. Self-powered electrospinning system driven by a triboelectric nanogenerator. ACS Nano. 2017;**11**(10):10439-10445. DOI: 10.1021/acsnano.7b05626

[53] Zi YL, Wu CS, Ding WB, Wang XF, Dai YJ, Cheng J, et al. Field emission of electrons powered by a triboelectric nanogenerator. Advanced Functional Materials. 2018;**28**(21):1800610. DOI: 10.1002/adfm.201800610

[54] Zhao K, Gu GQ, Zhang YN, Zhang B, Yang F, Zhao L, et al. The self-powered CO2 gas sensor based on gas discharge induced by triboelectric nanogenerator. Nano Energy. 2018;**53**:898-905. DOI: 10.1016/j.nanoen.2018.09.057

[55] Lv SS, Yu B, Huang T, Yu H, Wang HZ, Zhang QH, et al. Gas-enhanced triboelectric nanogenerator based on fully-enclosed structure for energy harvesting and sensing. Nano Energy. 2019;**55**:463-469. DOI: 10.1016/j.nanoen.2018.11.022

[56] Yang Y, Zhang HL, Wang ZL. Direct-current triboelectric generator. Advanced Functional Materials. 2014;**24**(24):3745-3750. DOI: 10.1002/adfm.201304295

[57] Luo JJ, Xu L, Tang W, Jiang T, Fan FR, Pang YK, et al. Direct-current triboelectric nanogenerator realized by air breakdown induced ionized air channel. Advanced Energy Materials. 2018;**8**(27):1800889. DOI: 10.1002/aenm.201800889

[58] Zhu XX, Li ZB, Li XS, Su L, Wei XY, Kuang SY, et al. Triboelectrification-enabled thin-film tactile matrix for self-powered high-resolution imaging. Nano Energy. 2018;**50**:497-503. DOI: 10.1016/j.nanoen.2018.05.061

[59] Ryu H, Lee JH, Khan U, Kwak SS, Hinchet R, Kim SW. Sustainable direct current powering a triboelectric nanogenerator via a novel asymmetrical design. Energy & Environmental Science. 2018;**11**(8):2057-2063. DOI: 10.1039/c8ee00188j

[60] Liu D, Yin X, Guo H, Zhou L, Li X, Zhang C, et al. A constant-current triboelectric nanogenerator arising from electrostatic breakdown. Science Advances. 2019;**5**(4):eaav6437. DOI: 10.1126/sciadv.aav6437

Chapter 5

Theoretical Prediction and Optimization Approach to Triboelectric Nanogenerator

He Zhang and Liwei Quan

Abstract

Triboelectric nanogenerator (TENG) is a new type of electrostatic generator based on the principle of Maxwell displacement current. It could be designed as a device for either smart sensing or energy harvesting via converting mechanical energy into electric power efficiently. To predict its output characteristic, investigate its working mechanism, and enhance its working performance, the theoretical analysis and optimization work in either experimental or theoretical means are of great significance. In this chapter, we plan to introduce the progress of theoretical analysis and optimization approach to TENG with four different modes. Three parts of work will be introduced in the manuscript: (1) the theoretical prediction approach for electric output performance of TENG device, (2) the optimization strategies for TENG device based on figure of merits, and (3) the scaling laws between the normalized electric outputs and multiple physical properties of the TENG device.

Keywords: triboelectric nanogenerator, theoretical analysis, scaling law, optimization strategy

1. Introduction

Triboelectric nanogenerator (TENG) [1, 2] is a revolutionary mechanical energy harvesting technology based on triboelectrification and induction effects of two materials with opposite electric polarities. This device contains two dissimilar dielectric films facing with each other, and there are electrodes deposited on the top and the bottom surfaces of the two films. The working mechanism of TENG is based on triboelectric effect and principle of displacement current. TENG may utilize the surface electrostatic charges produced in physical contact-separation process of the two insulators to generate electric power. By separating the tribo-pair with contact-induced triboelectric charges, a potential drop will be generated which forces the electrons to flow between the two electrodes. Compared with other renewable energy technologies including solar cells [3], biofuels [4], tide generator [5], etc., TENGs are less restricted by environmental or climatic conditions. In applications of mechanical energy harvesting, TENGs have much higher power output and energy conversion efficiency than those of electromagnetic [6] and piezoelectric energy harvesters [7–11]. Besides, TENGs can be made from low-cost materials by easy manufacture processes because of its simple device structures and configurations. These advantages make this new technology a better choice as power source for devices and microsystems [1, 2].

TENG has four basic operational modes with different structures: contact-separation mode, sliding mode, single-electrode mode, and freestanding mode. As **Figure 1** shows, these different modes are designed to meet the needs in different fields of applications. Based on these four modes, various TENG devices have been developed for different applications. Using these four modes and their combinations, a range of TENG devices have been invented for energy harvesting like wind and water, structure vibration and biomechanical motion energy harvesters, and self-powered smart sensing like vector sensor, tactile sensor, vibration detection, human physical signal detection, etc. [12–16].

With the progress of materials, structure design, and theories in fundamental mechanisms, the output performance of TENG is improved vigorously. Since its invention, the output power density of TENGs increases from initially a few $\mu W/m^2$ and mW/m^2 to tens of W/m^2. In recent works, a TENG with power density up to 500 W/m^2 and energy conversion efficiency more than 70% has been achieved, which is adequate to power most microelectronic devices and systems [2]. It is the optimization that makes the performance of the device greatly improved.

To further improve the output performance of the device, optimization design is of great importance and has already attracted attentions. In early works, the development of a new type of TENG devices or optimization of TENG design is typically realized through trial-and-error process experimentally, which is of high cost and time-consuming. Compared with experimental means, theoretical analysis is useful and more powerful in understanding the working mechanism of the device and could offer better optimization strategy for device structural design, material selection, and operation conditions. It is more convenient to realize the optimization of TENG with theoretical method rather than in experimental means. Based on the basic working mechanism of TENGs, the theoretical models for TENG have been established, a series of parameter analysis works have been carried out, and the relationship between the parameters and the output performance is obtained [17–21]. In addition, to establish a standard for evaluating different architectures

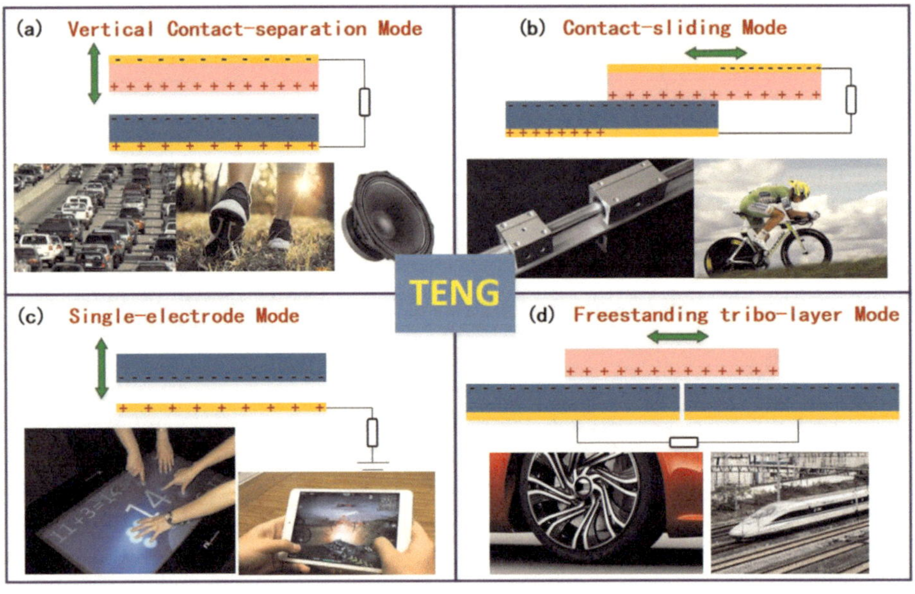

Figure 1.
Four fundamental modes of triboelectric nanogenerator [1].

of a TENG device, the device figure of merits of TENG is proposed and used to determine the maximum output power density of a TENG [22–24].

Although various theoretical models have been established for TENG, most of them are based on single-parameter analysis which focuses on investigating the effect of single variable on device performance with others fixed. For instance, in optimization of the thickness of dielectric materials, we may obtain its effect with other parameters fixed. The same procedure is required in repetition for other working conditions with different sets of fixed parameters. Yet the optimized structures and conditions may not be the best as many parameters are correlated.

To address this problem, Zhang proposed a set of formulations for normalized electrical outputs of the device in dimensionless forms [25, 26]. The expressions for these normalized outputs rely on two compound parameters composed of device dimensions (sizes, dielectric layer thickness), electrical properties of the electrode and dielectric materials, loading conditions (loading force, frequency, and motor process), and the circuit conditions (open/short circuit and load resistance). With these expressions, a multiparameter analysis method for theoretical approach for optimization of TENG has been derived. This method makes it possible to realize the optimization of the device by tuning different physical parameters simultaneously, which may reveal the real situation of the TENG that its output performance is influenced simultaneously and coherently by a number of factors.

In this chapter, the theoretical approaches of V-Q-x relationship for different TENGs are presented. Along with these relationships, the output performance of TENG and the optimization method with material and device figure of merits for TENG are discussed. Lastly, the multiparameter analysis method and the optimization strategy for TENG are presented.

2. The theoretical prediction and optimization strategies for TENG

In this section, the theoretical prediction for output performance and optimization strategies for design of TENG are discussed from three different aspects: the theoretical prediction approach for V-Q-x relationship of different modes of TENGs is presented firstly, followed by the principle of material and structure figure of merit for different modes of TENG and corresponding optimization strategies, and, lastly, the scaling laws between the normalized electric output of TENG and multiple physical properties are derived, with which the optimization strategies for TENG are provided.

2.1 The V-Q-x relationship of TENG

The four basic modes of TENG are of similar structure with two layers of different materials (dielectric-dielectric or dielectric-electrode), which are usually called tribo-pair. When the tribo-pair comes into contact, some charges move from one material to another to equalize their electrochemical potential due to triboelectric effect. When forced to separate, some of the surface charges tend to keep the original state, while the others tend to give electrons away, possibly producing triboelectric charges on the surfaces. The presence of triboelectric charges on dielectric surfaces can be a force for driving electrons in the electrode to flow in order to balance the electric potential drop created.

As **Figure 1** shows, there are two different motion patterns including contact-separation and sliding in these four modes of TENG. For the contact-separation mode and single-electrode mode TENG, the tribo-pair moves vertically and creates an air gap in between, while for the contact-sliding mode and freestanding tribo-layer mode, the tribo-pair moves laterally. For these two different kinds of motion process, the theoretical models are quite different, which will be discussed in Sections 2.1.1 and 2.1.2, respectively.

2.1.1 Contact-separation mode TENG

According to the materials used and device structures, the contact-separation mode TENGs fall into two categories: dielectric-to-dielectric contact and conductor-to-dielectric contact structures (**Figure 2**). Based on Gauss's law, we may consider this kind of TENG as a series of flat plate capacitors; the relationship between electric field E in the tribo-pair and the total charge density ρ is $\nabla \cdot E = \rho/\varepsilon_0$. When the charged surfaces move, according to Maxwell's equations, the electric displacement field is defined as $D = \varepsilon_0 E + P$; therefore, the current density is $J_D = \varepsilon_0 \frac{\partial E}{\partial t} + \frac{\partial P}{\partial t}$. Thus, we can get the V-Q-x relationship of TENG [18].

The thicknesses of the dielectric plates in tribo-pair are assumed to be d_1 and d_2 with the relative dielectric constants ε_{r1} and ε_{r2}, respectively. According to the Gauss theorem, the electric field strengths of each layer in the dielectric-to-dielectric mode TENG are $E_1 = -Q/S\varepsilon_0\varepsilon_{r1}$ for dielectric plate 1, $E_2 = -Q/S\varepsilon_0\varepsilon_{r2}$ for dielectric plate 2, and $E_{air} = (-Q/S + \sigma(t))/\varepsilon_0$ in the air gap, respectively. Here ε_0 is the vacuum permittivity, σ is the charge density at the contact surface of the tribo-pair, and S is the contact area of the tribo-pair. The voltage between the two electrodes is $V = E_1 d_1 + E_2 d_2 + E_{air} x(t)$. For the conductor-to-dielectric structure TENG, there will be no dielectric plate 2, i.e., d_2 and E_2 are zero; thus the voltage becomes $V = E_1 d_1 + E_{air} x(t)$. If we define the equivalent thickness of the dielectric plates to be $d_0 = d_1/\varepsilon_{r1} + d_2/\varepsilon_{r2}$, the V-Q-x relationship could be expressed as follows:

$$V = -\frac{Q}{S\varepsilon_0}(d_0 + x(t)) + \frac{\sigma x(t)}{\varepsilon_0} \tag{1}$$

Here, $x(t)$ is the varying gap distance between the tribo-pair due to external mechanical loading imposed to the TENG device. In special circuit conditions, for short-circuit condition with the output voltage $V = 0$, the current is $I_{sc} = \frac{\mathrm{d}Q}{\mathrm{d}t} = \frac{S\sigma d_0}{(d_0 + x(t))^2}\frac{\mathrm{d}x}{\mathrm{d}t}$; for open-circuit condition with transferred charge $Q = 0$, the voltage is $V_{oc} = \frac{\sigma x(t)}{\varepsilon_0}$.

According to Ohm's law, when the TENG is connected with a load resistance R to form a circuit, the output voltage in the circuit can be expressed as

$$V = IR = R\frac{\mathrm{d}Q}{\mathrm{d}t} \tag{2}$$

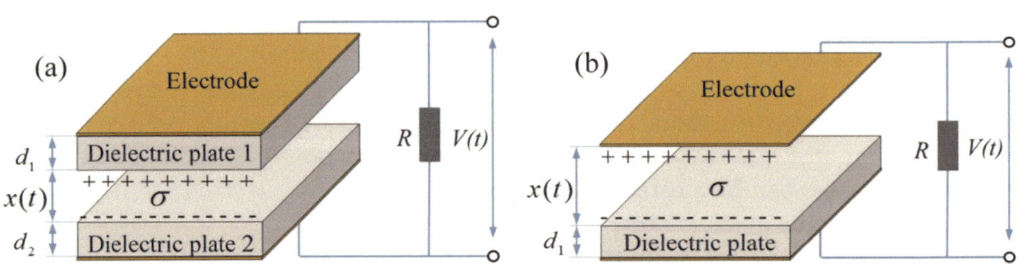

Figure 2.
Basic structure and model of the contact-mode TENG. (a) Dielectric-to-dielectric mode TENG and (b) conductor-to-dielectric mode TENG [18].

Substituting Eq. (1) into Eq. (2), we will have the equation for Q as

$$R\frac{\mathrm{d}Q}{\mathrm{d}t} = -\frac{Q}{S\varepsilon_0}(d_0 + x(t)) + \frac{\sigma x(t)}{\varepsilon_0} \tag{3}$$

With initial condition of $Q(t = 0) = 0$, the expression of $Q(t)$ will be

$$Q = \sigma S - \sigma S e^{-\frac{1}{RS\varepsilon_0}\left(d_0 t + \int_0^t x(t)\mathrm{d}t\right)} - \frac{\sigma}{R\varepsilon_0} e^{-\frac{1}{RS\varepsilon_0}\left(d_0 t + \int_0^t x(t)\mathrm{d}t\right)} \int_0^t d_0 e^{\frac{1}{RS\varepsilon_0}\left(d_0 z + \int_0^z x(t)\mathrm{d}z\right)} \mathrm{d}z \tag{4}$$

With the output voltage as

$$\begin{aligned} V(t) = R\frac{dQ}{dt} &= -\frac{\sigma d_0}{\varepsilon_0} + \frac{\sigma(d_0 + x(t))}{\varepsilon_0} \exp\left[-\frac{1}{RS\varepsilon_0}\left(d_0 t + \int_0^t x(t)\mathrm{d}t\right)\right] \\ &+ \frac{\sigma d_0}{\varepsilon_0} \frac{d_0 + x(t)}{RS\varepsilon_0} \int_0^t \exp\left(-\frac{d_0(t-\tau)}{RS\varepsilon_0} - \frac{1}{RS\varepsilon_0}\int_\tau^t x(z)\mathrm{d}z\right)\mathrm{d}\tau \end{aligned} \tag{5}$$

This V-Q-x relationship provides a means for evaluation of the electric output of contact-mode TENG, which may also be used for single-electrode mode TENG.

To verify the accuracy and precision of the proposed relationship, the experimental results from Ref. [16] are utilized here as an example for contact-mode TENG. Glass and polydimethylsiloxane (PDMS) are used as the tribo-pair in the experiment in planar form. The physical parameters of the device and experiment design are listed in **Table 1**. The TENG device is driven by a dynamic testing machine with separation gap $x(t)$ between tribo-pair (see **Figure 3a**, black solid line).

According to the experimental settings in Ref. [16], we introduce a piecewise function to describe the motion process of the tribo-pair as

$$x(t) = \begin{cases} 0, & 0 \le t \le (1-\eta)T, \\ A \sin\left(\dfrac{\pi t}{\eta T}\right), & (1-\eta)T < t \le T. \end{cases} \tag{6}$$

Here, η is the separation ratio representing the ratio of the separation time to the entire period. The smaller the separation ratio, the faster the tribo-pair gets into contact and thus the longer the contact time. In each period, the dielectric plates are supposed to start the charging process which occupies the time period from 0 to $(1-\eta)T < t \le T$ when the tribo-pair keeps contact ($x = 0$).

Parameter	Value
Thickness of PDMS plate d_1	100 μm
Relative permittivity of PDMS ε_{r1}	2.7
Thickness of glass d_2	1 mm
Relative permittivity of glass ε_{r2}	7.2
Permittivity of vacuum ε_0	8.854×10^{-12} F/m
Area of the dielectrics S	25 cm^2
Surface charge density σ	5–40 μC/m^2

Table 1.
Parameters of contact-mode TENG [16].

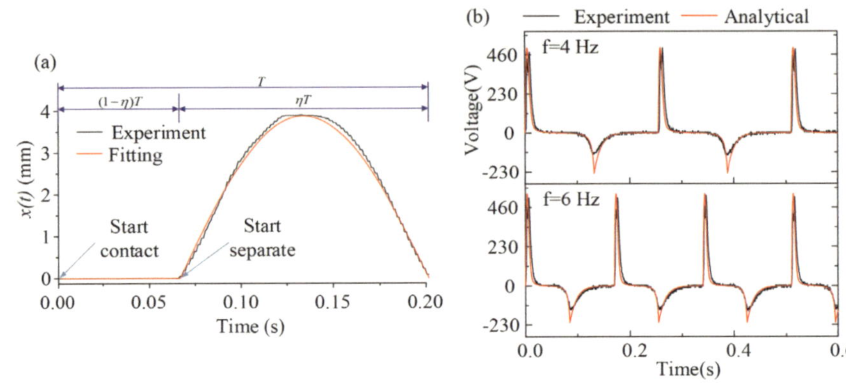

Figure 3.
(a) Oscillation scheme used in experiment (black line) and its fitting function that is input into the analytical solution (red line) and (b) comparisons of the theoretical predictions on output voltage with experimental measurements [25].

In **Figure 3b**, the voltage output obtained by Eq. (5) is compared with the experimental results from Ref. [16]. Excellent agreements could be observed between the experimental and analytical results due to different loading frequencies, especially the positive part of the voltage output. For the negative part of the results, the analytical results are slightly larger than the experimental ones, particularly for the low contact frequency case. The results presented above clearly indicate the accuracy of the model and method developed in this work.

2.1.2 Sliding-mode TENG

For lateral sliding-mode TENG, the basic structure is shown in **Figure 4**.

For the lateral sliding-mode TENG, the contact area is S, the sliding distance is $x(t)$, the length of the tribo-pair is l, and the width is w. Since l is always much larger than d_1 and d_2 as the tribo-pair, and x is always smaller than $0.9l$, it is difficult to keep tribo-pair in perfect alignment. In this condition, the total capacitance C can be estimated as $C = \varepsilon_0 w(l - x)/d_0$. Utilizing the charge distribution shown above and Gauss theorem, the V_{oc} can be estimated as $V_{oc} = \sigma x d_0/\varepsilon_0 w(l - x)$. So the V-Q-x relationship can be shown as

$$V = -\frac{Q}{C} + V_{oc} = -\frac{Qd_0}{\varepsilon_0 w(l - x)} + \frac{\sigma x d_0}{\varepsilon_0 w(l - x)} \tag{7}$$

For conductor-to-dielectric TENGs when the edge effect can be neglected, the relationship is similar by turning d_2 into zero. As a result, we can get the output voltage of sliding-mode TENG as

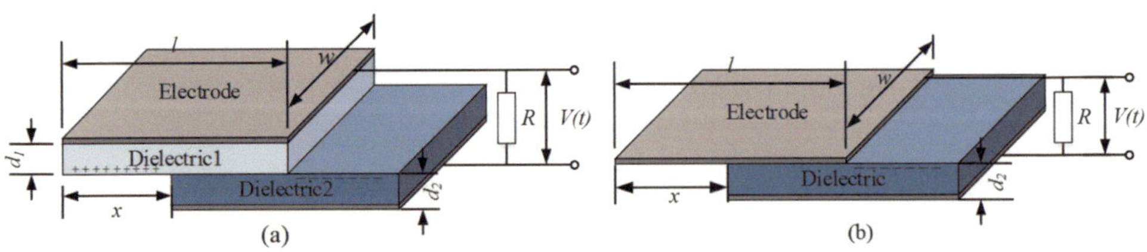

Figure 4.
Theoretical model of sliding-mode TENG. (a) Dielectric-to-dielectric sliding-mode TENG and (b) conductor-to-dielectric sliding-mode TENG [26].

$$V = \frac{\sigma d_0}{\varepsilon_0}\left[\frac{l}{l-x(t)}\exp\left(-\frac{d_0}{\varepsilon_0 RS}\int_0^t \frac{l}{l-x(t)}\mathrm{d}t'\right) + \frac{d_0}{\varepsilon_0 RS}\frac{l}{l-x(t)}\int_0^t \exp\left(\frac{d_0}{\varepsilon_0 RS}\int_t^{t'}\frac{l}{l-x(\delta)}d\delta\right)\mathrm{d}t' - 1\right] \quad (8)$$

Based on these output voltage expressions, we can get the average power output for TENG as

$$P_{\text{eff}} = \frac{1}{T}\int_t^{t+T}\frac{V^2}{R}\mathrm{d}t \quad (9)$$

Here, T is the time span of one mechanical loading cycle.

This V-Q-x relationship provides a method for evaluation of the electric output of sliding-mode TENG, which may be easily extended as a methodology for sliding freestanding tribo-layer mode TENG. To make sure the accuracy and precision of the proposed V-Q-x relationship of sliding-mode TENG, the experiments from Ref. [17] are carried out for validation. The materials and scale parameters are shown in **Table 2**.

In the experiments, the tribo-pair is fixed on a horizontal tensile loading platform for reciprocating lateral sliding process. The sliding process of the tribo-pair was imposed by the dynamic testing machine with a symmetric acceleration-deceleration mode (**Figure 5c**). The analytical results of the voltage output obtained by Eq. (8) agree very well with the experimental result.

Based on the V-Q-x relationships presented from Eqs. (5)–(8), the output performance of TENG is predictable when device structure, material parameters, and motion process are clear. With these equations, the influence of each parameter is clear, while all others are certain. For example, we can change the one parameter such as load resistance while all the others kept unchanged. As a result, we can get the output characteristic of the target TENG device and find out the optimized load resistance.

With this method, parametric analyses are carried out to characterize the output performance of TENGs with different working conditions. Niu et al. studied the output characteristics of contact-mode [18], sliding-mode [19], single-electrode

Parameter	**Value**
Dielectric tribo-pair	PTFE-nylon
Thickness of nylon plate d1 (μm)	50
Relative permittivity of Nylon εr1	4
Thickness of PTFE plate d2 (μm)	50
Relative permittivity of PTFE εr2	2.1
Permittivity of vacuum ε0 (F/m)	8.854×10^{-12}
Area of the dielectrics S (m^2)	0.05×0.071
Surface charge density σ (μCm^{-2})	200
Maximum separate distance A (m)	0.05
Acceleration a (m/s^2)	20
Load resistance R (MΩ)	10–1000

Table 2.
Parameters of sliding-mode TENG [17].

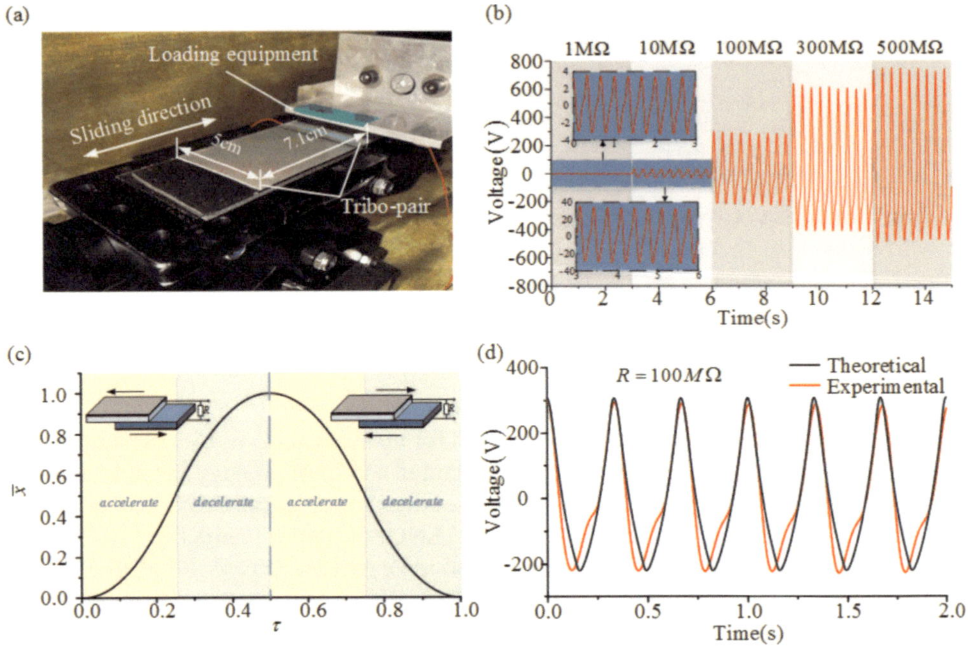

Figure 5.
(a) The TENG device in testing, (b) time history of output voltage, (c) simulation of dimensionless sliding process and (d) comparison of time history of output voltage between theoretical results and experimental results [26].

mode [20], and freestanding-mode [21] TENG under different load resistances, products of velocities, contact area sizes, effective dielectric thicknesses, and gap distances. These works obtained the effect of a series of independent parameters on the output characteristics including load resistance, maximum gap or slid distance, moving speed, device capacitance, and device structure parameters. These works, by numerical calculation of the real-time output characteristics, presented the suitable value range of these preceding parameters with common TENG device design and provided excellent guidance for structural design and optimization strategies for TENG devices.

2.2 Optimization approach based on figure of merit

For various applications, four basic modes of TENGs have been developed. Each mode has its own structure and various triboelectric material choices for the tribo-pair. That makes it difficult to characterize and compare the output performance of TENG. A universal standard has to be introduced to quantify the performance of the TENGs, regardless of its operation mode. To solve this problem, the figure of merit of TENG is proposed [22]. It gives a quantitative evaluation of TENG's performance from both structure's and the materials' points of view [22–24]. The application of figure of merit leads to a more efficient design and optimization approach of various TENG structures in practical applications. It may help to establish a series of standards for developing TENGs toward practical applications and industrialization.

The figure of merits of TENG includes performance figure of merit related to the structure and material figure of merit related to surface charge density. The theoretical derivation and simulation of these two figures of merit will be discussed in Sections 2.2.1 and 2.2.2, respectively.

2.2.1 Figure of merit for quantifying the output performance of TENG

Based on the V-Q-x relationship, a series of optimization strategies for independent parameters have been proposed. But their optimization target is the maximum output voltage or power, and the physical properties of the tribo-pair are not taken into account. To take into consideration the influence of other parameters of the device, the performance figure of merit (FOM_P) is developed as a new standard. It stands for the maximum power density of TENG, which represents a quantitative standard to reflect the output capability of TENGs with different configurations [22]. The expression of FOM_P is

$$FOM_P = 2\varepsilon_0 \frac{E_m}{Ax_{max}} \tag{10}$$

Here, E_m is the largest possible output energy per cycle, x_{max} is the maximum displacement of tribo-pair, and A is the contact area. The FOM_P is considered as a universal standard to evaluate varieties of TENGs, since it is directly proportional to the greatest possible average output power regardless of the mode and size of TENG [22]. To obtain the FOM_P, the power output per cycle of TENG is required first.

The output energy per cycle E can be calculated through the encircled area of the closed loop in the V-Q curve as

$$E = \overline{P}T = \int_0^T VI\mathrm{d}t = \int_{t=0}^{t=T} V\mathrm{d}Q = \oint V\mathrm{d}Q \tag{11}$$

The V-Q curve is usually obtained from the working process of TENG by the V-Q-x relation. In general, the working process of any type of TENG contains four steps as a cycle (**Figure 6a**). As **Figure 6b** shows, the V-Q curve becomes steady after a few initial cycles. The output energy per cycle derived from these steady periodic cycles is named as "cycles for energy output" (CEO).

For each CEO, the difference between the maximum and the minimum transferred charges in its steady state is defined as the total cycling charge Q_C. As **Figure 6b** shows, Q_C is always smaller than the maximum transferred charge $Q_{SC.max}$. If the Q_C could be maximized to $Q_{SC.max}$, we may achieve the maximum power output. Having noticed that the $Q_C = Q_{SC.max}$ only occurs in short-circuit condition, we add a switch in parallel with the external load and design the following repeated steps to achieve instantaneous short-circuit conditions during operations: step 1, the tribo-pair separates from $x = 0$ to $x = x_{max}$at switch off; step 2, turn switch on to enable $Q_C = Q_{SC.max}$; step 3, the tribo-pair displaces from $x = x_{max}$ to $x = 0$ at switch off; and step 4, turn the switch on to enable $Q_C = 0$. With this process, we may obtain the V-Q plot in **Figure 6c**. These cycles are named "cycles for maximized energy output" (CMEO) by Ref. [22].

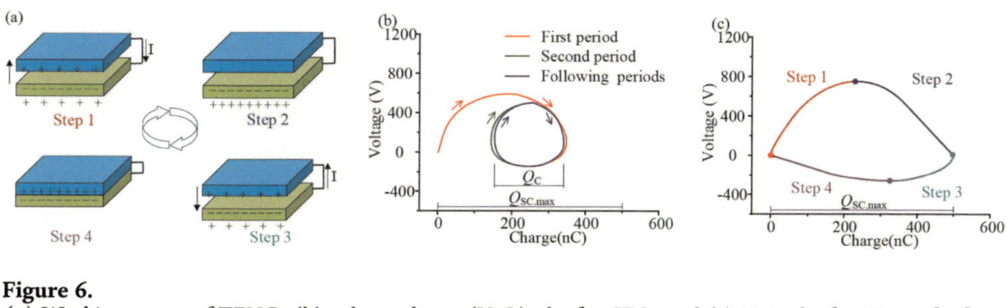

Figure 6.
(a) Working process of TENG, (b) voltage-charge (V-Q) plot for CEO, and (c) V-Q plot for CEMO [22].

The CMEO is related to load resistance of the output circuit: the higher the resistance, the higher the output energy per cycle. Therefore, the maximized output energy could be obtained when an infinitely large resistor as

$$E_m = \frac{1}{2} Q_{SC.max}\left(V_{OC.max} + V'_{max}\right) \tag{12}$$

Here $Q_{SC.max}$ is the short-circuit transferred charge, $V_{OC.max}$ is the maximum open-circuit voltage, and V'_{max} is the maximum achievable absolute voltage. The equation for E_m of TENG operated in any conditions can also be expressed as

$$E_m = \int_0^{t_1} P_1(t)dt + \int_{t_1}^{t_2} P_2(t)dt + \int_{t_2}^{t_3} P_3(t)dt + \int_{t_3}^{t_4} P_4(t)dt \tag{13}$$

The average power output $\overline{P}$ is then

$$\overline{P} = \frac{E_m}{T} \approx \frac{\overline{v} E_m}{2x_{max}} \tag{14}$$

Here $\overline{v}$ is the average velocity of the relative motion in tribo-pair. Thus, we can define the FOM_P depending on the parameters of E_m, x_{max}, and A and get its expression as Eq. (10).

2.2.2 Figure of merit for quantifying material characteristic

From Section 2.1, we notice that the transfer charge Q and output voltage V are proportional to the surface charge density σ in the V-Q-x relations of TENG. The increase of σ will directly enhance the possible average output power of TENG significantly as the power output is proportional to σ^2. Yet for FOM_P as Eq. (10) shows, the material characteristic is not taken into consideration. Thus we define the material figure of merit (FOM_m) as a standard to quantify the charge density of general surface as

$$FOM_m = \sigma^2 \left(C^2/m^4\right) \tag{15}$$

The FOM_m is only determined by σ and the component related to the material properties alone. It can be used to evaluate the triboelectric performance of the materials in contact [22].

The accurate value of σ for one kind of material is measured by putting it into a testing TENG device and using electrometer to measure the total transfer charge. To avoid the influence of uncertain contact conditions, the testing TENG device utilizes liquid metals as the other material of the tribo-pair to get the maximum possible surface charge of the material. Through this method, the FOM_m of a series of commonly used materials has been acquired, and their position in triboelectric series compared with the liquid metal has been figured out. These results in detail can be found in Ref. [22] and its following works.

2.2.3 Application of figure of merits

The figure of merits provides a series of quantitative standards to evaluate the working performance of TENG. Their application enables more efficient design and optimization of various TENGs in practical applications. The optimization works based on figure of merits are likely to establish the principles for TENG design and develop TENGs toward practical applications and industrialization.

Through FOM_P, we introduced a universal standard to quantify the power output performance of the TENG regardless of its operation mode and materials. With this standard, we are able to evaluate the performance of the TENGs in different structures/modes and achieve the optimization approach for structure design and working parameter setting. Using Eq. (11), we may obtain the FOM_P for different TENGs with same tribo-pair and contact area through analytical and simulation method. Taking the maximum FOM_P as an optimization index, we will achieve the optimized structure/mode for a TENG. On the other hand, for a known TENG, we can find out its maximum FOM_P and thus determine the best suitable working condition like load resistance, x_{max}, etc. Through these analyses based on FOM_P, the following conclusions have been found: (1) x_{max} influences the output of TENG directly, and when x_{max} grows, the total transferred charges will increase and so does the output power; (2) the power output performance of TENG triggered by contact-separation action is better than that of sliding action; and (3) reducing the parasitic capacitance can help to increase the power output of TENG, and also larger-area generators with larger air gap capacitance are much more robust to parasitic leakage [22–24].

Using FOM_m, we have introduced a quantitative standard to evaluate the maximum surface charge density for a triboelectric material. With FOM_m of different materials, we are able to determine their triboelectric charge polarity. And with different FOM_m of tribo-pairs, we can find the optimal choice for a TENG. With this standard, we found some useful regulations: (1) the optimized choice of materials in a tribo-pair should be in opposite polarities, and (2) the larger the polarity differential between the materials, the bigger the FOM_m of the tribo-pair [22].

2.3 Optimization strategy for TENG based on multiparameter analysis

For TENGs, their electric output performance is simultaneously and coherently influenced by a group of factors including the dimensions of the electrodes and insulators, electrical properties of the materials and the loading processing, etc. These parameters are interlinked with each other; therefore changing one parameter with the others fixed may break the optimized condition of the device and require new adjustment in other parameters and further optimization. It is necessary to simulate the output performance of a TENG via theoretical models based on multiparameter analysis rather than specific cases with only one variable considered. In addition, TENGs could be used as sensor or energy harvester in either macro- or microscale, within which the physical properties of materials and device are quite different. Hence, the dimensionless analysis method is more feasible for parameter analysis of TENG.

Here, to realize the optimization of the device, we developed a series of normalized expressions for output voltage and output power in dimensionless forms, which provide a group of scaling laws between the normalized electric output and two independent compound variables. These scaling laws can facilitate the analysis of the effects in different aspects of the device simultaneously and provide accordance for optimal design of TENG by considering the effects of all factors simultaneously. The optimal electric output could be obtained through the proposed formulations with all parameters of the TENG considered as variables [25, 26].

2.3.1 Contact-separation mode TENG

For the optimization of contact-separation mode TENG, a set of expressions for normalized electric voltage and power in dimensionless forms are proposed in this section. In these dimensionless expressions, the effects of all the parameters involved have been investigated comprehensively and simultaneously. The normalized expressions for output voltage and power may facilitate the optimization based on the scaling laws by tuning different physical properties simultaneously rather than those only focusing on one physical property either in dimensional or dimensionless forms [25].

For the output voltage in a general circuit with the load resistance of R, the various structure, material, mechanical loading, and circuit properties of TENG are investigated in a coherent manner and expressed with two combined parameters. The dimensionless expression for normalized output voltage that depends only on the two combined parameters is derived based on Eq. (5) as

$$\frac{V(\tau)\varepsilon_0}{\sigma A} = \overline{V}_A\left(\tau, \frac{A}{d_0}, \frac{RS\varepsilon_0}{AT}\right) = -\frac{d_0}{A} + \left(\frac{d_0}{A} + \overline{x}(\tau)\right)\exp\left[-\frac{AT}{RS\varepsilon_0}\left(\frac{d_0}{A}\tau + \int_0^\tau \overline{x}(\tau)d\tau\right)\right]$$
$$+\frac{AT}{RS\varepsilon_0}\frac{d_0}{A}\left(\frac{d_0}{A} + \overline{x}(\tau)\right)\int_0^\tau \exp\left[\frac{AT}{RS\varepsilon_0}\left((\zeta - \tau)\frac{d_0}{A} + \int_\tau^\zeta \overline{x}(\zeta)d\zeta\right)\right]d\zeta \tag{16}$$

where $\tau = t/T$ is the dimensionless time and A and T are, respectively, the oscillation amplitude and the period of the separation-contact cycle.

While for the application of energy harvest, the output power should be a key variable for characterizing the performance of the generator. According to the definition of output power $P_{\mathrm{eff}} = \frac{1}{R}\sqrt{\frac{\int_0^T V^2(t)dt}{T}}$, the dimensionless output power may be given by

$$\frac{P_{\mathrm{eff}} T\varepsilon_0}{\sigma^2 AS} = \overline{P}_A\left(\frac{A}{d_0}, \frac{RS\varepsilon_0}{AT}\right) \tag{17}$$

Here, we can see that the dimensionless output voltage and power depend only on two combined parameters, i.e., A/d_0 and $RS\varepsilon_0/AT$, in which A/d_0 characterizes the relative oscillation amplitude, while $RS\varepsilon_0/AT$ reflects the hybrid effects of generator area, electrical resistance, oscillation amplitude, and period of the tribo-pair. Besides this universal optimization of TENG by tuning multiple parameters at the same time, we may still realize the optimization for individual physical quantity with the assumption of other parameters fixed. However, we cannot figure out how to optimize the oscillation amplitude A to achieve the peak output from Eqs. (16) and (17) since this physical property get involved in both the combined dimensionless parameter and the normalized electric output. For the same reason, we may not work out the effect of the generator area S and the mechanical loading period T on the output power with Eq. (17).

If we want to examine the effect of S on the output power, we may simply multiply Eq. (17) with the factor $RS\varepsilon_0/AT$ and have

$$\frac{P_{\mathrm{eff}} R\varepsilon_0^2}{\sigma^2 A^2} = \overline{P}_{AR}\left(\frac{A}{d_0}, \frac{RS\varepsilon_0}{AT}\right) \tag{18}$$

To investigate the effect of oscillation amplitude and period, another set of dimensionless expressions is derived by multiplying Eqs. (16) and (17) with the factor A/d_0 and the combined parameter $RS\varepsilon_0/AT$ and multiplying Eq. (18) with A^2/d_0^2 and $RS\varepsilon_0/AT$. This set of dimensionless expressions for the scaling laws will be

$$\frac{V(\tau)\varepsilon_0}{\sigma d_0} = \overline{V}_d\left(\frac{A}{d_0}, \frac{RS\varepsilon_0}{d_0 T}\right) = -1 + \left(1 + \frac{A}{d_0}\overline{x}(\tau)\right)\exp\left[-\frac{d_0 T}{RS\varepsilon_0}\left(\tau + \frac{A}{d_0}\int_0^\tau \overline{x}(\tau)d\tau\right)\right]$$
$$+\frac{d_0 T}{RS\varepsilon_0}\left(1 + \frac{A}{d_0}\overline{x}(\tau)\right)\int_0^\tau \exp\left[\frac{d_0 T}{RS\varepsilon_0}\left((\zeta - \tau) + \frac{A}{d_0}\int_\tau^\zeta \overline{x}(\zeta)d\zeta\right)\right]d\zeta \tag{19}$$

$$\frac{P_{\text{eff}} T\varepsilon_0}{\sigma^2 d_0 S} = \overline{P}_d\left(\frac{A}{d_0}, \frac{RS\varepsilon_0}{d_0 T}\right) \text{ or } \frac{P_{\text{eff}} R\varepsilon_0^2}{\sigma^2 d_0^2} = \overline{P}_{dR}\left(\frac{A}{d_0}, \frac{RS\varepsilon_0}{d_0 T}\right) \tag{20}$$

The relations in Eqs. (16)–(20) also reveal that the effective output voltage (or power) is proportional to the square of surface charge density. These dimensionless parameters and corresponding scaling laws also show a straightforward optimization method for the magnitude of oscillation at the same electrical load resistance, generator area, or oscillation period.

To verify the dimensionless expressions, we compare the peak dimensionless output voltage based on Eqs. (16) and (19) with the experimental results in corresponding dimensionless forms. Two groups of experiments with various thicknesses d_{glass} of glass plate [16] are utilized for comparisons. In Group I, the oscillation frequency $1/T$ varies from 1 Hz to 6 Hz with the load resistance fixed at $R = 100\ M\Omega$, while in Group II, the load resistance R varies from 1 MΩ to 1000 MΩ with the oscillation frequency fixed at $1/T = 5$ Hz. The theoretical predictions for the dimensionless voltage $V_{\text{peak}}\varepsilon_0/\sigma A$ and $V_{\text{peak}}\varepsilon_0/\sigma d_0$ (solid lines) both agree very well with experiments (markers) in **Figure 6**.

As can be seen from **Figure 7**, the theoretical predictions agree very well with the experimental measurements despite that the latter are achieved with different device structures, mechanical loadings, and circuit conditions. This demonstrates that the established scaling law reveals the underlying general correlation between the physical properties of the device and its output performance, which may provide robust guidelines for optimization strategies, no matter what way we use to make it, theoretical or experimental. Take the varying parameters T and R in the two experimental groups as an instance: from the experimental point of view, they are totally different parameters, of which the variations affect the output performance, respectively. However, we find in the scaling law the unified expression for the two parameters in $RS\varepsilon_0/AT$ (or $RS\varepsilon_0/d_0T$). The individual variation of either T or R has since become the variation of the compound variables. That is to say, the same compound variable may have different combinations of T and R. According to the scaling laws shown in Eqs. (16)–(20), the output voltages should have the same value in the cases with the same compound variables $RS\varepsilon_0/AT$ (or $RS\varepsilon_0/d_0T$) and A/d_0 even if the latter are comprised of parameters valued with different combinations. It indicates the necessity for coupling design of these parameters included in the compound variables to achieve the best output performance of the device.

However, the correlations among the parameters and their simultaneous effect on the output performance are hard to be found in traditional optimization techniques based on single-parameter investigations, in neither experimental nor theoretical means. As thus, we can conclude that the scaling law proposed in this paper can be not only used to predict the output performance of a TENG comprehensively and systematically with all parameters being considered simultaneously but also treated as a general and rational optimization criterion for the device toward its best performance. Based on the scaling laws from Eqs. 16 to 20, the output performance of the generator can be optimized by tuning combined parameters or individual physical quantities. The scaling law for the output voltage and power in a form of either in Eqs. (16)–(18) or in Eqs. (19)–(20) provide a universal optimizing strategy

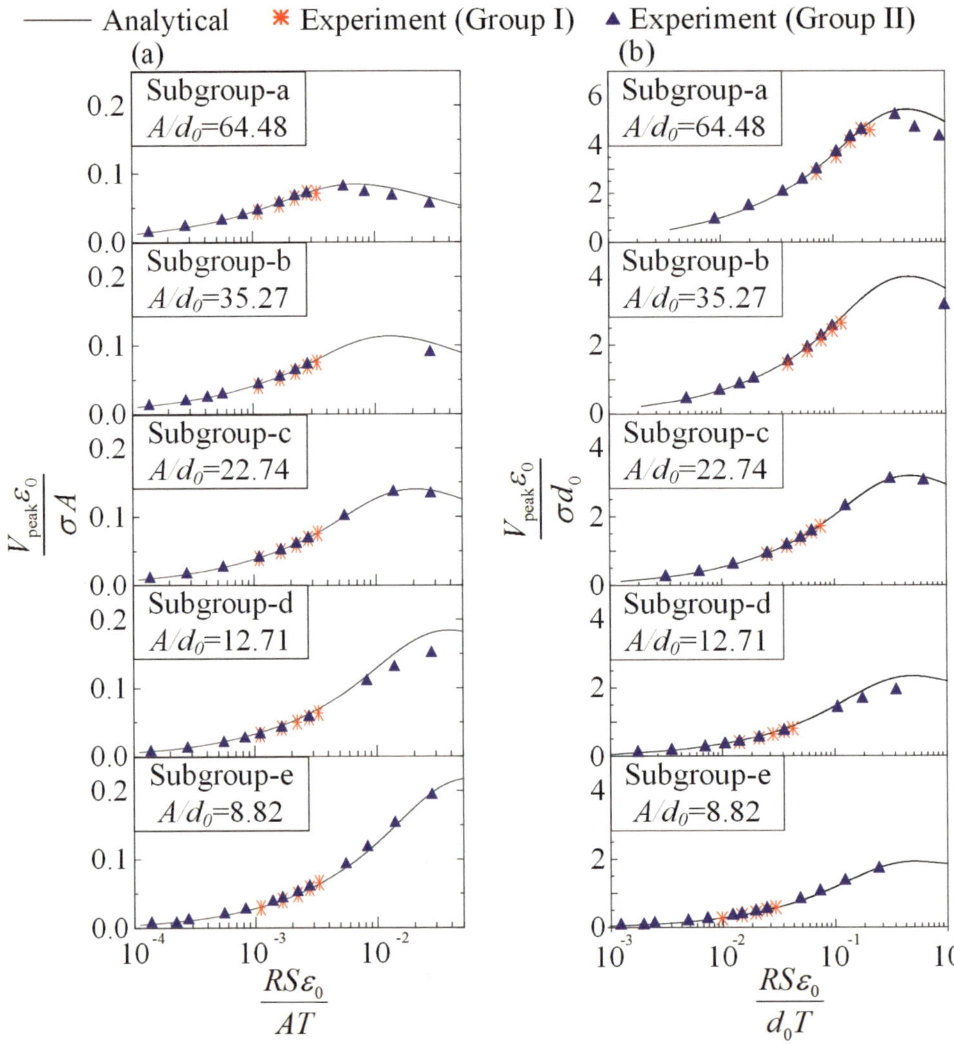

Figure 7.
Validation of the scaling laws for dimensionless peak output voltage through comparisons with experimental measurements with different setups [25].

to enhance the output voltage for sensing application or maximum output power for energy harvesting application [25].

In **Figure 8**, the maximum peak voltage is obviously increasing monotonically with the relative oscillation amplitude A/d_0 (**Figure 8a**). While for each given value of A/d_0, there exists an optimized $RS\varepsilon_0/AT$ that delivers a maximum dimensionless output power (**Figure 8b**). It suggests that we may achieve the best output performance of TENG through tuning multiple physical parameters simultaneously. From **Figure 8**, we can also find that the pinpoint for a peak output power or power density (=0.25) is located around $A/d_0 = 2.0$ and $RS\varepsilon_0/AT = 0.12$.

Besides the universal optimization with multiple parameters, we may also use individual parameter analysis by tuning its quantity to enhance the output performance. For instance, from **Figure 8b** it is found that given all other parameters, there exists an optimal load resistance to achieve a maximum output power. While from **Figure 8c** with given load resistance and oscillation amplitude, it can be found that larger generator area and higher oscillation frequency may enhance the output power until saturated.

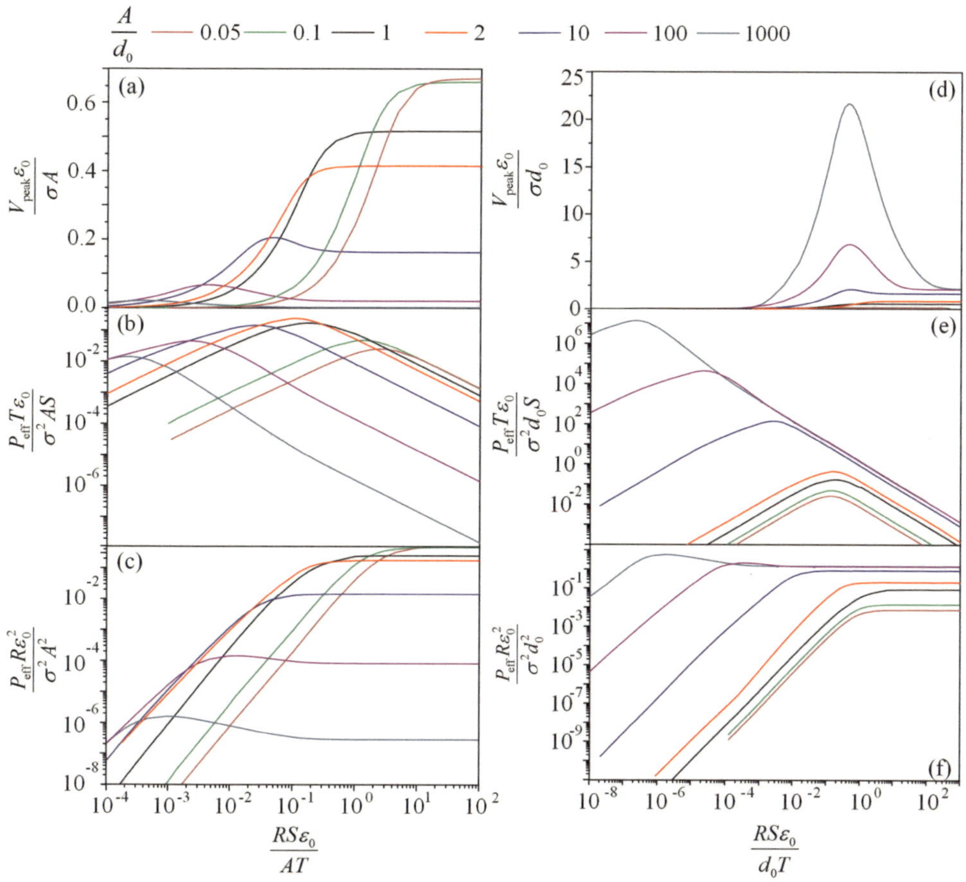

Figure 8.
Scaling laws for peak output voltage and output power [25].

As mentioned before, the scaling laws in Eqs. (16)–(19) cannot reflect a clear dependence of the output performance on the oscillation amplitude A since it is included in both two dimensionless parameters. Fortunately, with the scaling laws in Eqs. (20) and (21), the effect of A is extracted and can be distinguished quantitatively from the numerical results. **Figure 8e–f** shows that increasing the oscillation amplitude may enhance the output power or power density monotonically and only a smaller generator area, lower oscillation frequency, or load resistance may achieve an optimized output power [25].

2.3.2 Sliding-mode TENG

For the optimization of sliding-mode TENG, Zhang [26] established a series of dimensionless expressions of output performance of sliding-mode TENG according to Eq. (7) as

$$\overline{V}_R\left(\tau,\frac{A}{l},\frac{\varepsilon_0 RS}{d_0 T}\right)=\frac{\varepsilon_0 V(\tau)}{\sigma d_0}=\frac{1}{1-\frac{A}{l}\overline{x}(\tau)}\exp\left(-\frac{d_0 T}{\varepsilon_0 RS}\int_0^{\tau}\frac{1}{1-\frac{A}{l}\overline{x}(\tau')}\tau'\right)$$
$$+\frac{d_0 T}{\varepsilon_0 RS}\frac{1}{1-\frac{A}{l}\overline{x}(\tau)}\int_0^{\tau}\exp\left(\frac{d_0 T}{\varepsilon_0 RS}\int_{\tau}^{\tau'}\frac{1}{1-\frac{A}{l}\overline{x}(\delta)}d\delta\right)d\tau'-1 \tag{21}$$

Here, A is the maximum sliding distance. When used as energy harvesters, the output power will become the key target to evaluate the output characteristics of TENG. According to the definition of effective output power, the dimensionless output power can be expressed as

$$\overline{P}_T\left(\frac{A}{l},\frac{\varepsilon_0 RS}{d_0 T}\right)=\frac{\varepsilon_0^2 RP_{\text{eff}}}{\sigma^2 d_0^2}=\int_0^1 \overline{V}^2 d\tau \tag{22}$$

In Eqs. (21) and (22), there are two dimensionless compound parameters A/l and $\varepsilon_0 RS/d_0 T$ affecting the dimensionless output characteristics of device, which could be described with a group of parameters related to various aspects of TENG device. From these equations, we may understand the effects of the load resistance R, planer area S, loading period T, and maximum sliding distance A on output voltage at the same time. Similarly, based on Eq. (23), we can figure out the optimized parameters for power improvement. Meanwhile, the dimensionless expressions also reflect the relationship between the compound parameters (A/l and $\varepsilon_0 RS/d_0 T$) and the output performance.

To make the normalized output performance better understood from the physical point of view, they can also be described with the dimensionless capacitance $\overline{C}$ and the dimensionless time constant $\overline{T}$, in which the expression for $\overline{C}$ is $\overline{C}=C_A/C_0=1-A/l$, with $C_0=\varepsilon_0 S/d_0$ the capacitance when $x=0$ and $C_A=C(x=A)=\varepsilon_0 S(1-A/l)/d_0$ the equivalent capacitance of the device when $x=A$. Here $\overline{C}$, from the physical point of view, represents the ratio of the capacitance with $x=A$ to the capacitance with $x=0$. The dimensionless time constant is defined as $\overline{T}=RC_0/T=\varepsilon_0 RS/d_0 T$, which reflects the time constant for the first-order circuits with $C=C_0$ to the period of the alternating current. The expression for normalized output voltage versus $\overline{C}$ and $\overline{T}$ is

$$\begin{aligned}\overline{V}_R(\tau,\overline{C},\overline{T})&=\frac{\varepsilon_0 V(\tau)}{\sigma d_0}=\frac{1}{1-(1-\overline{C})\overline{x}(\tau)}\exp\left(-\frac{1}{\overline{T}}\int_0^\tau\frac{1}{1-(1-\overline{C})\overline{x}(\tau')}d\tau'\right)\\&+\frac{1}{\overline{T}}\frac{1}{1-(1-\overline{C})\overline{x}(\tau)}\int_0^\tau\exp\left(\frac{1}{\overline{T}}\int_\tau^{\tau'}\frac{1}{1-(1-\overline{C})\overline{x}(\delta)}d\delta\right)d\tau'-1\end{aligned} \tag{23}$$

$$\overline{P}_T(\overline{C},\overline{T})=\frac{\varepsilon_0^2 RP_{\text{eff}}}{\sigma^2 d_0^2}=\int_0^1 \overline{V}_R^2(\tau)d\tau \tag{24}$$

Similar to Eqs. (16)–(18) of contact-mode TENG, in this circumstance, the influence of d_0 on normalized output voltage is not reflected in Eq. (23), neither the effects of d_0 and R on normalized output power from Eq. (24). To find out the influence of these parameters, we propose the following set of expressions for normalized output voltage and power equations as

$$\begin{aligned}\overline{V}_d(\tau,\overline{C},\overline{T})&=\frac{TV(\tau)}{\sigma RS}=\frac{1}{\overline{T}}\frac{1}{1-(1-\overline{C})\overline{x}(\tau)}\exp\left(-\frac{1}{\overline{T}}\int_0^\tau\frac{1}{1-(1-\overline{C})\overline{x}(\tau')}d\tau'\right)\\&+\frac{1}{\overline{T}^2}\frac{1}{1-(1-\overline{C})\overline{x}(\tau)}\int_0^\tau\exp\left(\frac{1}{\overline{T}}\int_\tau^{\tau'}\frac{1}{1-(1-\overline{C})\overline{x}(\delta)}d\delta\right)d\tau'-\frac{1}{\overline{T}}\end{aligned} \tag{25}$$

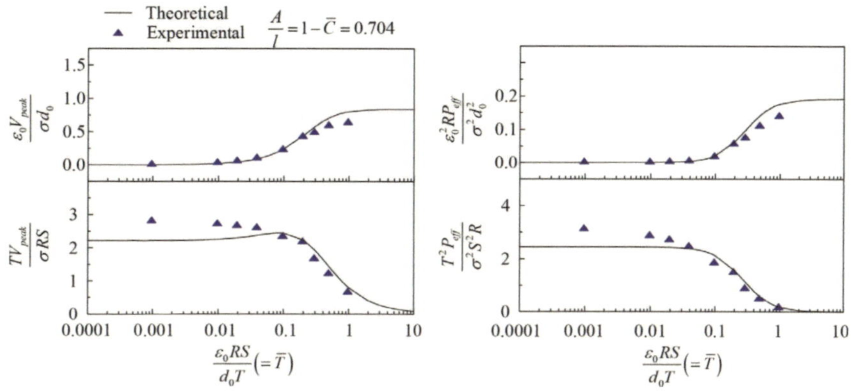

Figure 9.
Comparisons of dimensionless voltage and power between theoretical results and experimental results [26].

$$\overline{P}_d(\overline{C},\overline{T}) = \frac{T^2 P_{\text{eff}}}{\sigma^2 S^2 R} = \frac{1}{\overline{T}^2}\int_0^1 \overline{V}_R^2(\tau)d\tau \tag{26}$$

$$\overline{P}_R(\overline{C},\overline{T}) = \frac{\varepsilon_0 T P_{\text{eff}}}{\sigma^2 d_0 S} = \frac{1}{\overline{T}}\int_0^1 \overline{V}_R^2(\tau)d\tau \tag{27}$$

These equations provide the optimal strategy for TENG design with the two compound parameters related to the TENG device. We may use these equations to study the physical interpretation of dimensionless expressions through the dimensionless capacitance and the dimensionless time constant.

To verify the dimensionless expressions, we compare the theoretical results with the experimental measurements using the peak output voltage in dimensionless forms [17]. The corresponding parameters used in experiment and simulation are listed in **Table 2**.

As **Figure 9** shows, compared with the experimental results, the theoretical curves based on scaling laws exhibit good consistent tendency with the experimental data points.

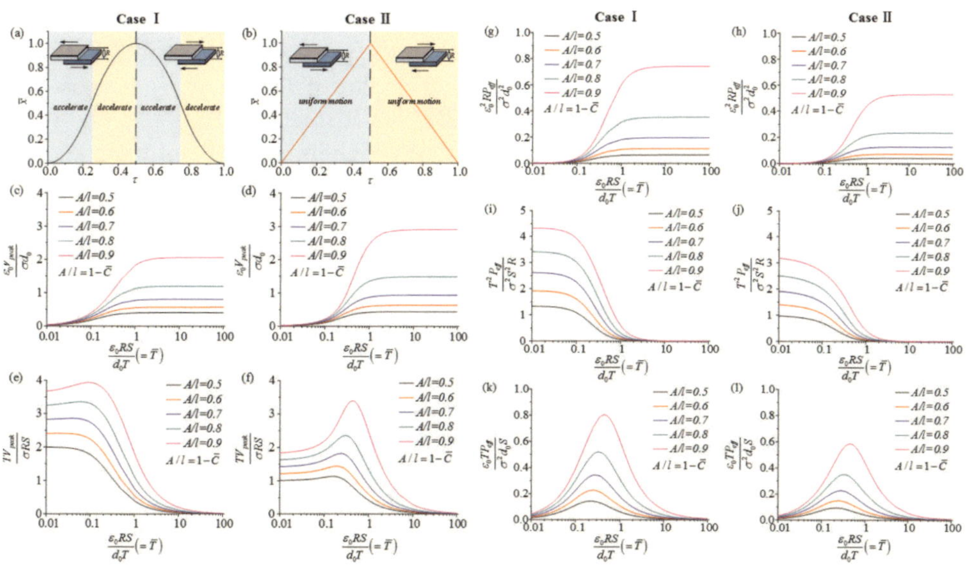

Figure 10.
Scaling laws for dimensionless output voltage and power in case I and case II [26].

According to the referring experiments [17], we define the two different forms of sliding process to simulate different mechanical loading conditions, including parabolic (Case I) and triangular (Case II) loading pattern. Based on the scaling laws for the correlation between the normalized output performance and the dimensionless compound parameters, some general optimization strategies will be acquired to improve the electric output with Eqs. (23)–(27).

From **Figure 10**, the optimal combination of A/l and $\varepsilon_0 RS/d_0 T$ for the best output performance for TENG is found, thus providing a general guideline for device optimization. In addition, we can improve the dimensionless output performance by tuning compound parameters as well as single parameter. We can also use the scaling law from another physical point of view by reflecting the output performance with the dimensionless capacitance $\overline{C}$ and time constant $\overline{T}$. The value of $\overline{C}$ and $\overline{T}$ with regard to the best output performance provides the optimization strategy for dimensionless capacitance and time constant.

From **Figure 10**, it is also obvious that increasing A/l may enhance both the peak output voltage and the maximum output power of TENG. It indicates that the larger the ratio of the maximum sliding distance to the length of the device, the higher the output performance is of the TENG. Additionally, it is found that the optimization strategies for TENG with these two different loading patterns are similar because the scaling laws between output performance and compound parameters in Cases I and II show the same trend.

In **Figure 10(c–d)**, for each A/l, the value of $\varepsilon_0 V_{peak}/\sigma d_0$ increases monotonically with $\varepsilon_0 RS/d_0 T$ and approaches to a constant when $\varepsilon_0 RS/d_0 T = 1.0$, which implies that the optimal combination $\varepsilon_0 RS/d_0 T$ is over 1. As mentioned before, the scaling laws shown in **Figure 10(c–d)** may not reflect the explicit relationships between output voltage and equivalent thickness d_0, because d_0 is included in both $\varepsilon_0 V_{peak}/\sigma d_0$ and $\varepsilon_0 RS/d_0 T$. To achieve the influence of d_0, another scaling laws are exhibited in **Figure 10(e–f)** with Eqs. (23) and (25). Instead of d_0 in **Figure 10(c–d)**, load resistance R, area S, and period T in **Figure 10(e–f)** are involved in both $TV_{peak}/\sigma RS$ and $\varepsilon_0 RS/d_0 T$, which means the effect of R, S, and T on output voltage may not be reflected in **Figure 10(d–f)**. For each A/l, a peak value of $TV_{peak}/\sigma RS$ is found, which indicates the existence of the optimal $\varepsilon_0 RS/d_0 T$ or $\overline{T}$.

In **Figure 10(g–h)** for the relationship between the dimensionless output power and compound parameters, the value of $\varepsilon_0^2 RP_{eff}/\sigma^2 d_0^2$ increases with $\varepsilon_0 RS/d_0 T$ firstly and then reaches a constant. At the same time, the value of $T^2 P_{eff}/\sigma^2 S^2 R$ decreases with $\varepsilon_0 RS/d_0 T$ after being a constant when $\varepsilon_0 RS/d_0 T$ is smaller than 0.1. In **Figure 10 (k–l)**, a peak value is observed for $\varepsilon_0 TP_{eff}/\sigma^2 d_0 S$, which implies the optimal parameter combination for $\varepsilon_0 RS/d_0 T$ or $\overline{T}$. Similar to output voltage, output power may also be enhanced by tuning single parameter. For example, the remaining area S, period T, and equivalent thickness d_0 fixed, the best combination of $\varepsilon_0 RS/d_0 T$and A/l can be obtained for optimizing the output power for TENG by tuning load resistance R [26].

According to **Figures 8** and **10**, we can conclude that the scaling laws can provide a more comprehensive and rational optimization strategy for both contact-separation mode and sliding-mode TENG based on multiparameter analysis. The results can help enhance the output performance of the device as either a smart sensor or an energy harvester and may render a guideline for designing TENG devices.

Acknowledgements

This work was supported by the National Key R&D Program of China under grant No. 2018YFB1600200 and National Natural Science Foundation of China under grant Nos. 11472244, 11621062, and 11772295 and the Fundamental Research Funds for the Central Universities under grant No. 2019QNA4040.

Author details

He Zhang* and Liwei Quan
College of Civil Engineering and Architecture, Zhejiang University, Hangzhou, China

*Address all correspondence to: zjuzhanghe@zju.edu.cn

References

[1] Wang ZL. Triboelectric nanogenerators as new energy technology and self-powered sensors—Principles, problems and perspectives. Faraday Discussions. 2014;**176**:447-458

[2] Wang ZL. On Maxwell's displacement current for energy and sensors: The origin of nanogenerators. Materials Today. 2017;**20**:74-82

[3] Oregan B, Gratzel M. A low-cost, high-efficiency solar cell based on dye-sensitized colloidal TiO_2 films. Nature. 1991;**353**:737-740

[4] Brennan L, Renew PO. Biofuels from microalgae—A review of technologies for production, processing, and extractions of biofuels and co-products. Sustainable Energy Reviews. 2010;**14**: 557-577

[5] Mago PJ, Blunden LS, Bahaj AS. Performance analysis of different working fluids for use in organic Rankine cycles. Proceedings of the Institution of Mechanical Engineers, Part A: Journal of Power and Energy. 2007;**221**:255-264

[6] Li Z, Zuo L, Luhrs G, Lin L, Qin Y. Electromagnetic energy-harvesting shock absorbers: Design, modeling, and road tests. IEEE Transactions on Vehicular Technology. 2013;**62**(3): 1065-1074

[7] Zhang H, Shen MZ, Zhang YY, Chen YS, Lu CF. Identification of static loading conditions using piezoelectric sensor arrays. ASME Journal of Applied Mechanics. 2018;**85**:011008-011005

[8] Chen YS, Zhang H, Zhang ZC, Lu CF. Theoretical assessment on piezoelectric energy harvesting in smart self-powered asphalt pavements. Journal of Vibration Engineering & Technologies. 2018;**6**(1):1-10

[9] Chen YS, Zhang H, Zhang YY, Li CH, Yang Q, Zheng HY, et al. Mechanical energy harvesting from road pavements under vehicular load using embedded piezoelectric elements. Journal of Applied Mechanics-Transactions of the ASME. 2016;**83**(8)

[10] Lu CF, Zhang YY, Zhang H, Zhang ZC, Shen MZ, Chen YS. Generalized optimization method for energy conversion and storage efficiency of nanoscale flexible piezoelectric energy harvesters. Energy Conversion and Management. 2018;**182**:34-40

[11] Zhang H, Ye GR, Zhang ZC. Acoustic radiation of a cylindrical piezoelectric power transformer. ASME Journal of Applied Mechanics. 2013;**80** (6) 061019(1-6)

[12] Yang Y, Zhang H, Lin ZH, Zhou YS, Jing Q, Su Y, et al. Human skin based triboelectric nanogenerators for harvesting biomechanical energy and as self-powered active tactile sensor system. ACS Nano. 2013;**7**(10): 9213-9222

[13] Ya Y, Hulin Z, Qingshen J, Yusheng Z, Xiaonan W, Zhonglin W. ACS Nano. 2013;7(8):7342-7351

[14] Changbao H, Chi Z, Xiaohui L, Limin Z, Tao Z, Weiguo H, et al. Self-powered velocity and trajectory tracking sensor array made of planar triboelectric nanogenerator pixels. Nano Energy. 2014;**9**:325-333

[15] Zhang H, Zhang JW, Hu ZW, Quan LW, Shi L, Chen JK, et al. Waist-wearable wireless respiration sensor based on triboelectric effect. Nano Energy. 2016;**59**:75-83

[16] Chen JK, Guo HW, Ding P, Pan RZ, Wang WB, Xuan WP, et al. Nano Energy. 2016;**30**:235-241

[17] Niu S, Liu Y, Wang S, Lin L, Zhou YS, Hu Y, et al. Theory of sliding-mode triboelectric nanogenerators. Advanced Materials. 2013;**25**(43):6184-6193

[18] Niu SM, Wang SH, Lin L, Liu Y, Zhou SY, Hu YF, et al. Theoretical study of contact-mode triboelectric nanogenerators as an effective power source. Energy & Environmental Science. 2013;**6**:3576

[19] Simiao N, Sihong W, Ying L, Shengyu Z, Long L, Youfan H, et al. A theoretical study of grating structured triboelectric nanogenerators. Energy & Environmental Science. 2014;**7**:2339

[20] Simiao N, Ying L, Sihong W, Long L, Yusheng Z, Youfan H, et al. Theoretical investigation and structural optimization of single-electrode triboelectric nanogenerators. Advanced Functional Materials. 2014;**24**: 3332-3340

[21] Simiao N, Yin L, Xiangyu C, Sihong W, Yusheng Z, Long L, et al. Theory of freestanding triboelectric-layer-based nanogenerators. Nano Energy. 2015;**12**: 760-774

[22] Yunlong Z, Simiao N, Jie W, Zhen W, Wei T, Zhonglin W. Standards and figure-of-merits for quantifying the performance of triboelectric nanogenerators. Nature Communications. 2015;**6**:8376

[23] Jiajia S, Tao J, Wei T, Xiangyu C, Liang X, Zhonglin W. Structural figure-of-merits of triboelectric nanogenerators at powering loads. Nano Energy. 2018;**51**:688-697

[24] Peng J, Kang SD, Snyder GJ. Optimization principles and the figure of merit for triboelectric generators. Science Advances. 2017;**3**:eaap 857615

[25] Zhang H, Quan LW, Chen JK, Xu CK, Zhang CH, Dong SR, et al. A general optimization approach for contact-separation triboelectric nanogenerator. Nano Energy. 2019;**56**:700-707

[26] Zhang H, Zhang CH, Zhang JW, Quan LW, Huang HY, Jiang JQ, et al. A theoretical approach for optimizing sliding-mode triboelectric nanogenerator based on multi-parameter analysis. Nano Energy. 2019. DOI: 10.1016/j.nanoen.2019.04.057